Success in Math

Basic Geometry

Teacher's Resource Manual

Executive Editor: Barbara Levadi
Market Manager: Sandra Hutchison
Senior Editor: Francie Holder
Editors: Karen Bernhaut, Douglas Falk, Amy Jolin
Editorial Assistant: Kris Shepos-Salvatore
Educational Consultant: Kathleen Coleman
Production Manager: Penny Gibson
Production Editor: Walt Niedner
Interior Design: The Wheetley Company
Electronic Page Production: The Wheetley Company
Cover Design: Pat Smythe

Printed in the United States of America 1 2 3 4 5 6 7 8 9 10 99 98 97 96

ISBN 0-8359-1192-6

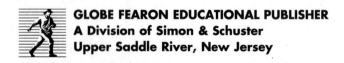
GLOBE FEARON EDUCATIONAL PUBLISHER
A Division of Simon & Schuster
Upper Saddle River, New Jersey

Contents

English Proficiency Strategies

Mathematics is a language consisting of carefully defined symbols that represent fundamental concepts. Students learning English as a second language (ESL) or limited in English proficiency (LEP) may need more time to make connections between concepts and mathematical language. All students in mathematics classes benefit from communicating their thinking, using mathematical language. Students who are developing proficiency with English mathematical language need opportunities to listen, speak, read, and write.

Teachers are encouraged to draw upon what students already know conceptually and should not assume that students understand mathematical terms. It is important for teachers to use new terms frequently in the context of class instruction and discussion and to encourage students to define mathematical terms in their own words. By using symbols that are common to basic mathematics operations, teachers help ESL students ease into English.

Here are some techniques for working effectively with ESL/LEP students in *Basic Geometry*.

1. Use hands-on activities with manipulatives, including tangrams and geoboards. Two- and three-dimensional shapes model geometric concepts and can be linked to terminology. Boxes and containers can be used to demonstrate geometric properties such as surface area or volume. Use tools such as compasses, protractors, and rulers for geometric construction and measurement.

2. Collect visual cues—pictures of geometric structures, diagrams, and drawings.

3. Students can create visual displays and label them. They can make drawings of arcs, angles, and line segments on index cards, with the corresponding terms or symbols on the reverse side.

4. Create opportunities to respond to and use examples of mathematics from other cultures. Encourage all students to make up their own problems, enabling them to cast their stories in familiar cultural perspectives.

5. Some students may be learning a new system of weights and measures for volume, length, time, temperature, and money. Hands-on practice may help them to overcome difficulties. Encourage students to make comparisons with the system of measurement used in their native country.

6. Build in reinforcement with repetition, paraphrasing, restatement, and the use of synonyms. Vary discussion techniques by including short questions and answers that provide a change of pace and reduce the pressure of more complex sentence construction. During question-and-answer times, allow enough wait time for students to respond so they can decode the English words and perform computations.

7. Take time to introduce words that have more than one meaning, such as *square*—a polygon with four equal sides and four 90-degree angles, and to multiply a number by itself.

8. Plan cooperative/collaborative activities, using structured techniques that combine small group interactions with individual accountability. Sometimes pairing an ESL student with an English-proficient speaker in a peer tutoring interaction increases opportunities for understanding.

9. To give students practice reading application problems, use group activities in which students circle the names of shapes or underline the words for mathematics operations.

10. Monitor student progress by reviewing work frequently and adjusting instruction to address student understanding.

11. Provide bilingual dictionaries, glossaries, and visual materials for classroom reference.

Cooperative Learning

Cooperative learning activities provide an alternative to traditional whole-class instruction and individual paper-and-pencil activities. Past practice emphasized individual computation and right answers. Systematic use of small groups for cooperative learning encourages a community of active learners working together. Students can help each other learn the material and extend their mathematical knowledge.

Small groups provide structures for students to exchange ideas, ask questions, and clarify concepts. Since mathematics problems can often be solved by more than one approach, students can exchange problem-solving strategies. As students explain their reasoning, they develop their use of mathematical language.

Cooperative learning is more than putting students together in small groups and giving them a mathematical task to solve. It involves careful attention to group process. The teacher can take an active role in structuring the learning situation by

- placing students in random or heterogeneous groups.
- providing structures for team building to encourage student collaboration.
- practicing appropriate social skills, such as learning how to give encouragement and constructive feedback.
- rearranging the classroom so that groups have their own workspace.
- outlining the problem and learning objectives for students; letting them know what they are going to do, why they are doing it, and how they are expected to work.
- providing appropriate materials.

There are different ways to structure cooperative learning groups. Some teachers start mathematics class with students in groups to check homework. The expectation is that each student will have attempted each problem. Students are expected to deal with each other's questions and reach consensus on solutions.

Other teachers use small groups as a follow-up to whole-group instruction. Each group works on problems related to the lesson. Some cooperative learning strategies suggest that teachers assign roles to group members. There are a number of leadership and management functions, such as recording information and tying ideas together by presenting conclusions.

Lessons best suited for cooperative learning have many possible answers or solution strategies. During cooperative learning time, the teacher circulates from group to group, observing student interaction, offering assistance where needed, and asking questions to keep groups working productively. Individual accountability is fostered by establishing a pattern requiring that each student attempt to solve a problem. Students compare results and resolve differences. As teachers circulate, they can keep in touch with how students are responding to the mathematics content.

Time for summarizing is an important part of cooperative learning. As students share solutions and questions, the discussion can help them to generalize from a specific problem by looking for patterns or relationships in the data.

Newcomers to cooperative learning may find it helpful to work with a teacher who has been using the technique. Some school districts have formed support groups for sharing questions or problems.

Balanced approaches—whole group, small group, and individual—offer a variety of situations for students. A well-planned cooperative learning situation engages students in *actively* doing mathematics.

Manipulatives

The importance of manipulatives has been documented for students at the elementary level, but manipulatives have not been widely used to teach mathematics at the secondary level. Mathematics at the high school level moves to abstract and symbolic thinking, beyond the concrete thinking represented by manipulatives. Developmentally, high school students are expected to have reached a level of abstract and symbolic thinking beyond the need for concrete experiences.

Recently, the standards set forth by the National Council of Teachers of Mathematics have recognized the need for actively involving students in the development and application of mathematical concepts. They recognize that some students may need opportunities for informal activities and the use of concrete materials as they work toward a greater level of abstraction. Teachers can structure students' work with manipulatives toward that end.

Manipulatives provide multisensory learning experiences that allow students to model concepts and to move beyond specific examples to generalizations. The following are some tips to guide your use of manipulatives.

1. Choose a material that embodies the concept you want to teach. No manipulative is adaptable to every situation.

2. Allow students time for independent exploration of a new material and plan activities that allow students to focus on the concept at hand.

3. As much as possible, provide materials for students to use rather than rely on teacher demonstration.

4. Use manipulatives in small groups or paired working situations. Students learn from others as they work together.

5. Encourage students to verbalize and discuss ideas when they work with models.

6. Provide more than one material that demonstrates a concept. As students experience concepts in more than one way, they move toward generalization and abstraction.

7. Model the use of appropriate mathematical language as you discuss students' ideas and questions.

8. As students explore a concept, help them connect what they are doing to ways of recording the results and to the use of appropriate symbols.

9. Remember that manipulatives do not teach by themselves. Materials, other students, carefully selected tasks, and teacher guidance are all important parts of the learning process.

Some manipulatives that work particularly well with *Basic Geometry* are:

- **Geoboards** These tools allow students to construct two-dimensional polygons, investigate area and perimeter, work with similar and congruent figures, and transform one figure into another. Activities can be extended with the use of geodot paper for recording and problem-solving.

- **Tangrams** Students can use tangrams to demonstrate isosceles triangles, congruence and similarity, area and perimeter, and ratio.

- **Three-dimensional geometric shapes** These manipulatives are useful for illustrating geometric properties.

- **Paper strips with brads or pushpins** Students can use these materials to construct angles, arcs, and rays.

- **Compasses and protractors** Geometric constructions made with these tools provide opportunities for sharing ideas and acquiring mathematical language.

Problem Solving

Recent reports in mathematics research have suggeted problem solving as the principal reason for studying mathematics. Reports from the workplace stress the need for workers who can use their mathematics skills and work collaboratively to solve problems. Our students need opportunities to develop problem-solving strengths.

We recognize the need to develop students' abilities to use a variety of strategies and techniques for solving problems. It is important to stress the *process* of problem solving. Students might learn how to solve particular problems when solutions are stressed. They are *more likely* to learn how to approach and solve other problems when process skills are also emphasized.

Problem solving usually includes these steps.

- Understanding the question.
- Finding the needed information.
- Planning what to do.
- Carrying out the plan.
- Checking the answers.

This model can help students see that problem solving requires several actions. Small groups can provide the setting for students rephrasing problems in their own words and clarifying their understanding of the question. Similarly, they can work through the other steps with ample opportunities to share their ideas and revise or expand their thinking.

This model provides an overview of steps involved in problem solving, but it does not suggest strategies or techniques for carrying out the steps. Here are some generalizations for problem-solving strategies.

- Problem-solving strategies can be taught. They help students explore possible solutions.
- No one strategy fits all problem situations. Some problems require more than one strategy.
- Students should have a number of strategies from which to choose. Encourage students to solve different problems with the same strategy, and the same problem with different strategies.

Some strategies for solving problems are:

- **Act it out.** This strategy helps students visualize what is involved in the problem. They go through the actions using manipulatives (or themselves) to make clearer the relationships among the parts of the problem.

- **Make a drawing or diagram.** This strategy provides a way to depict the information to make relationships apparent. Stress with students that elaborate drawings are not necessary. They should draw only what is essential to tell about the problem.

- **Look for a pattern.** The focus is on relationships between elements. Students may make a table and use it to find a pattern.

- **Construct a table.** This strategy can be useful for organizing data that helps students find a pattern or missing information.

- **Guess and check.** Students may have been discouraged from guessing in the past, but this can be a valuable strategy. Encourage students to make knowledgeable guesses and emphasize the need to check the answers. Reassure students that many guesses may be necessary and remind them that there may be more than one solution.

- **Work backwards.** Sometimes problems state the final conditions and ask about something that took place earlier.

- **Identify wanted, given, and needed information.** This stategy is useful in understanding the problem and the planning stages of the problem-solving model.

The process of checking answers is important for developing strong problem-solving skills. During this step, students look back at how they solved the problem, giving them the opportunity to discuss, clarify, or revise their thinking. Students generalize from one situation to another. Teachers can build on this process by asking students to find another way to solve the problem or find another solution.

Together, these approaches help students gain confidence in their ability to become mathematical problem solvers.

Chapter 1 Introduction to Basic Elements

Introduction
Point out specific uses ancient civilizations had for geometry, such as mapping cities, navigating the Mediterranean Sea, building structures such as the pyramids, and even the movement of armies. Then ask students how geometry might be applied today. They may discover many comparable uses.

Lesson 1•1 Points, Lines, Planes

Alternate Teaching Approach
Model points, lines, and planes with a street map and straight pins. Identify several locations on the map and then have students work in pairs or small groups to find other locations. After leading a discussion of the concepts of points, lines, and planes, ask students to show other points, lines, and planes by using any part of the map. An intersection can be a point, a street can be a line, and the boundaries of the city could define a plane.

Working with ESL/LEP Students
Have students describe points, lines, and planes in their own words. Encourage them to think of as many different examples as possible. They could draw pictures of their ideas or find examples in magazines. Then have students work in small groups to make three posters, one showing examples of points, one of lines, and one of planes.

Final Check
To be sure students understand the concepts of point, line, and plane, discuss the following questions.

1. Can a line contain four points? Why or why not? [Yes, a line contains many points.]

2. Can a point that is in a plane be on a line that is in the plane? Draw a picture to prove your answer. [Yes, if a line is in a plane, all points on that line are in the plane.]

3. Can a plane contain four different lines? Draw a picture to prove your answer. [Yes, drawings should show four coplanar points with each pair of points connected—six different connections will be possible. Each of these six lines lies in the plane.]

Lesson 1•2 Line Segments and Rays

Using Manipulatives
Use pins and pieces of string to demonstrate line segments and rays. Give students time to experiment with these materials in small groups. Ask them to complete the following activities.

1. Use a pin to represent point A. Use a piece of string to represent a ray with endpoint A. If you do not move point A, how many different rays have endpoint A? [many different rays] Make a drawing of your work and label the rays. [Drawings will look like a sunburst with point A at the center.]

2. Use a second pin to represent point X. Use a piece of string to represent a line segment from endpoint X to endpoint A. If you do not move point X, how many different line segments can have both endpoints? [one] Make drawings of your work and label the line segments.

Alternate Teaching Approach
Ask students what they think when they hear the words *segment* and *ray*. For example, students may think of an orange segment or a ray of sunshine. Use their ideas to lead into a discussion of the lesson.

Error Analysis
Be sure students understand the differences between line segments and rays. Emphasize that a ray has only one endpoint, while a line segment has two endpoints.

Lesson 1•3 Angles

Chalkboard Activity
For review and to lead into this lesson, students should draw each figure described below.

1. Two rays with the same endpoint

2. Ray CD and ray CE

3. Two rays with endpoint R

Using Manipulatives
Model an angle with a pushpin and two paper strips. Pin the end of each strip to one point. Show students how to use these materials to form different angles. Pass out these materials and encourage small groups to experiment with them. Each group could write questions about the angles they make. Groups might exchange and answer one another's questions.

Error Analysis

When naming angles, be sure students know that the vertex must be the middle letter. Encourage students who struggle with this concept to think of the two rays (the paper strips) that form the sides of the angle. Remind them that the endpoint of both rays is the same but should be named only once.

Lesson 1•4 Parallel and Perpendicular Lines

Using Manipulatives

Invite students to use two pens, pencils, or pieces of uncooked spaghetti to investigate the relationships between two lines. As students work with these materials, ask them to record any thoughts they have about the relationships. Encourage students to share their ideas with the class as you begin discussion of this lesson.

Working with ESL/LEP Students

Encourage students to describe the concepts of parallel lines and perpendicular lines in their own words. Draw the symbols for parallel and perpendicular on the board. If necessary, students could be encouraged to define these terms in their primary language on index cards, with the symbols next to them. They can refer to these terms when necessary. Encourage them to add cards with other new terms and symbols throughout the year.

Final Check

To be sure students understand the concepts of parallel and perpendicular lines, ask them to draw three lines. Two of the lines should be parallel and the third line should be perpendicular to the other two lines.

Lesson 1•5 Angles Formed by Intersecting Lines

Chalkboard Activity

Have students draw figures, using the directions described below, and answer the questions.

1. Draw two intersecting lines that are not perpendicular. How many angles are formed? [4]

2. Draw two parallel lines. Draw a third line that intersects the parallel lines but is not perpendicular to them. How many angles are formed? [8]

Using Manipulatives

Model intersecting lines with three pens, pencils, or pieces of uncooked spaghetti. Ask students to find the different relationships that can exist among three lines. As students work, suggest they write down their thoughts about the relationships. Have them think about their answers to the Chalkboard Activity as they finish. Ask students to share some ideas with the class as you begin discussion of this lesson.

Working with ESL/LEP Students

Explain terms with drawings, pictures, and/or hand motions whenever possible. Ask students to define each term in their own words. Encourage them to draw as many examples as needed to grasp the concepts and then add the drawings to their index card files. Encourage students to discuss their ideas with one another.

Lesson 1•6 More Pairs of Angles

Chalkboard Activity

Have students draw and answer the following.

1. Draw $\angle ABC$ and $\angle CBD$. What do these angles have in common? [$\overrightarrow{BC}$]

2. Draw a line. Draw points X, Y, and Z on the line. Draw ray YW What angles are formed? [$\angle XYW$ and $\angle WYZ$]

3. Draw $\overrightarrow{MN}$. Let point M be the vertex. Draw two angles that have ray MN as a side.

Using Manipulatives

Use a clock with movable hands, including a second hand, to demonstrate the concept of adjacent angles. Give small groups of students their own clocks for studying adjacent angles and linear pairs.

Alternate Teaching Approach

If students cannot identify or draw adjacent angles, then encourage them to draw and cut out rays to manipulate. Ask them to arrange the rays to form two angles on their desktops so that one ray from the first angle overlaps a ray of the second angle. Point out that the names of adjacent angles always contain two of the same letters.

Error Analysis

Be sure students understand that adjacent angles are side by side and always share one side or ray.

Lesson 1•7   Intersecting and Parallel Planes

Chalkboard Activity

Challenge students to draw figures, using the following directions.

1. Draw a plane. Draw a line in the plane. Can you draw a different plane that contains the line? Why or why not? [Yes, because there are many different planes that contain the line]

2. Draw two parallel lines. Can you draw two planes that contain the lines? Why or why not? [Yes, because all lines are part of a plane]

Using Manipulatives

After discussing the lesson, have small groups of students use pieces of paper or cardboard to show intersecting planes, parallel planes, skew lines, and a half plane. As students work, have them record their thoughts about the relationships.

Working with ESL/LEP Students

Use the ideas below to help students understand this lesson.

1. Relate the concepts of intersecting and parallel planes to the concepts of intersecting and parallel lines.

2. Relate the meaning of the term *half* to the meaning of the term *half plane*.

3. Encourage students to use two pens or pencils to demonstrate many different pairs of skew lines.

Chapter 1 Test

Use the figure at the right to answer exercises 1 to 15. .

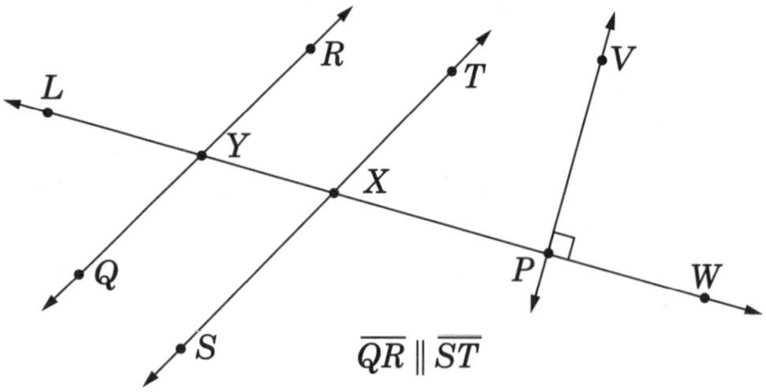

1. Name three points. _____
2. Name two line segments.

   _____
3. Name two lines. _____
4. Name two rays. _____
5. Name a pair of adjacent angles. _____
6. Name a pair of vertical angles. _____
7. Name a linear pair of angles. _____
8. Name a right angle. _____
9. Name a pair of perpendicular lines. _____
10. Name a pair of parallel lines. _____
11. Name a transversal. _____
12. Name a pair of alternate interior angles. _____
13. Name a pair of alternate exterior angles. _____
14. Name three collinear points. _____
15. Name a pair of intersecting lines. _____

$\overline{QR} \parallel \overline{ST}$

Solve these problems.

16. What advantages are there in making the roof of a structure parallel to its floor?

   _____

17. The floors of a building are not perpendicular to the walls; what can this tell you about the structure?

   _____

18. What kind of angle does an open door make with the wall in which it is set?

   _____

19. Draw how intersecting railroad tracks might look.

20. Draw pictures to show your answers to exercises 16 and 17.

Chapter 2 Angles

Introduction

Let students demonstrate the angles formed by one object bouncing off another. For example, they could roll a tennis ball into a wall at a certain angle and notice the angle formed when the ball bounces away. The same demonstration can be done with light from a flashlight reflected at an angle in a mirror.

Lesson 2•1 Classifying Angles

Chalkboard Activity

Ask students to draw at least four different angles on the board. Then have them compare the size of each angle with the corner of a book or a sheet of paper. Ask whether each angle is smaller or larger than the corner. Then place a straightedge so that it lines up with one side of the corner. Have students use the straightedge and corner to form two 90° angles and then state whether or not each student angle has a side that falls between the corner and the straightedge.

Using Manipulatives

To model angles, use a pushpin to anchor four thin strips of paper at one end. Give these materials to small groups of students. Challenge them to set three strips to form a right angle and a straight angle. Ask groups to use the fourth strip to form several acute angles. Remind them to record their findings. Then ask groups to try using the fourth strip to form several obtuse angles. Have groups share their findings with the whole class.

Working with ESL/LEP Students

Have students use their own words to write definitions on note cards. For example, *straight* is a word most students know; this helps them remember the meaning of *straight angle*. For terms such as *acute* and *obtuse*, encourage students to draw pictures to help them remember the meanings.

Error Analysis

If students confuse the terminology for types of angles, more practice finding and classifying different angles may help. Encourage students to find magazine pictures showing different angles and to make posters featuring each type of angle.

Lesson 2•2 Recognizing Complementary and Supplementary Angles

Chalkboard Activity

Read the following instructions and questions to students as they draw the figures.

1. Draw a right angle. Then draw a ray inside the right angle.

   a. How many angles are inside the right angle?
   [2]

   b. What is the sum of their measures?
   [90°]

2. Draw a straight angle. Then draw a ray inside the straight angle.

   a. How many angles are inside the straight angle?
   [2]

   b. What is the sum of their measures?
   [180°]

Alternate Teaching Approach

To model complementary and supplementary angles, use a pushpin to anchor three thin paper strips at one end. Give these materials to small groups of students. Have them use the strips to form two complementary angles. Ask students to describe the position of the strips. [Students should find that two of the strips make a right angle and the third strip is between these two.] Ask students to form two supplementary angles and describe the position of the strips. [Students should find that two strips make a straight angle and the third strip is between these two.]

Using a Calculator

Find the measure of an angle that is complementary or supplementary to another angle, as in exercises 12 and 13. Remind students that storing *90* or *180* in the calculator memory simplifies repeated work.

Error Analysis

Students may need time to review terms and symbols for labeling angles correctly. Remind them to use three letters, with the middle letter showing the vertex of the angle. Practice matching letters with angle diagrams may help.

Lesson 2•3 Using a Protractor to Measure Angles

Using Manipulatives

Students can make a tool to help them line up the sides of angles on a protractor. Give students two thin strips of cardboard and a brad. The strips form the sides of an angle with the brad anchoring the vertex. Make sure the strips are long enough to reach the angle measurement lines. As students become more comfortable measuring angles, encourage them to use a circular or semi-circular protractor.

Encourage students to use a protractor to measure angles formed by objects in the classroom. Examples might include the legs of a chair or desk, the slanting cover of a three-ring binder, and the angle formed by a partially open door. Challenge them to guess the measure of the angle before using the protractor.

Error Analysis

Students who incorrectly measure angles may be improperly aligning the angle with the protractor. Be sure students place the vertex of the angle at the center point of the protractor and then align one of the sides with the line labeled *0* on the protractor. They then should read the scale starting with *0*.

Final Check

Have students draw at least four angles of different measures. Ask them to measure one another's angles. To review concepts taught earlier in this chapter, ask students to classify each angle as right, straight, acute, or obtuse, and to note any angles that are supplementary or complementary.

Lesson 2•4 Using a Protractor to Draw Angles

Chalkboard Activity

With students working in small groups, ask each member to draw four angles on the board. Have students measure one another's angles. Then discuss ways to draw angles with specified measures. Ask each group to share its ideas with the class.

Error Analysis

When checking students' angle measures for accuracy, note answers that are supplements of the correct measure. This shows that students are measuring properly, but using the incorrect protractor scale to measure the angle. Remind students that acute angles are between 0° and 90° and obtuse angles are between 90° and 180°. It may be helpful for students to draw one ray and then draw several angles with different measures, each having that ray as a side.

Final Check

Ask students to draw four rays with different endpoints that point in four different directions. Then have students exchange papers and use the rays to draw angles with measures of 60°, 120°, 40°, and 165°. Remind them to label each angle. Display the papers so students can see and check the work of others.

Lesson 2•5 Recognizing Congruent Line Segments and Angles

Chalkboard Activity

Ask students to draw figures, using the following directions, and to answer the questions.

1. Draw two line segments with the same length. Must the segments be parallel? Why or why not?
   [No, the line segments can have the same length and intersect or never touch.]

2. Draw two angles with the same measure. Must the rays point in the same direction? Why or why not?
   [No, the angles can have the same measure no matter what direction their rays point.]

Using Manipulatives

Give students models for identifying line segments and angles. Tangrams and geometric shapes such as triangles, squares, rectangles, and pentagons are useful. Have students find as many congruent line segments and angles as possible in each figure. Encourage students to trace the figures onto a sheet of paper and then label all corresponding parts.

Error Analysis

The Chalkboard Activity shows students that congruence is not affected by orientation. Those having difficulty with this concept should be encouraged to repeat the Chalkboard Activity until they gain confidence. If necessary, draw several pairs of line segments and angles to help them start.

Lesson 2•6 Bisecting Line Segments

Chalkboard Activity
Use the following activities to introduce midpoint.

1. Draw two congruent line segments.

2. Draw two congruent line segments with a common endpoint.

3. Draw two congruent line segments that lie on the same line.

4. Draw two line segments that have all the properties in steps 1 through 3.

Working with ESL/LEP Students
Remind students that *mid-* means *middle* and that *bi-* means *two*. Relate these definitions to those of *midpoint* and *bisect*. Encourage students to write these definitions in their own words on index cards and to draw pictures that show the word meanings.

Error Analysis
Students may need to review terms and symbols in order to use them with confidence. Encourage students to keep a list of terms and symbols to which they can refer and review from time to time. Some examples in this lesson are *line segment, arc, intersects, midpoint, bisect,* and the symbol for congruence ($\cong$). Students might make a chart illustrating these symbols.

Lesson 2•7 Bisecting Angles

Chalkboard Activity
Use the following activities to introduce bisecting angles.

1. Draw an angle with a measure of 80°.

   a. What is the measure of an angle half this size? [40°] Use the drawing of an 80° angle to draw it.

   b. Describe the ray that forms the angle in part a. [It divides the angle into two equal parts or bisects it.]

2. Repeat Activity 1 with a 120° angle. [a. 60°. Two 60° angles should be created by this ray. b. It divides the angle into two equal parts or bisects it.]

Using Manipulatives
Duplicate the situation described at the beginning of the lesson by using a circular piece of cardboard to represent a pizza. Or, have a pizza party. Have students experiment with slices of different sizes and record their work.

Error Analysis
If students confuse bisecting angles with complementary and supplementary angles, have volunteers draw and bisect two angles, 170° and 100°, on the chalkboard. Working in groups, students can find supplementary, complementary, and adjacent angles. Ask them to share their conclusions with the whole class.

Lesson 2•8 Using Parallel Lines

Chalkboard Activity
Students should complete the following to review parallel lines and transversals and to discover congruent angles formed by two parallel lines cut by a transversal.

1. Draw two parallel lines.

2. Draw two lines and a transversal.

3. Draw two parallel lines and a transversal.

Measure each pair of alternate interior angles. How are they related? [Each pair has the same measure.]

Using Manipulatives
Use a children's gate or an expandable rack like the one below to illustrate the concepts in this lesson. Ask students to identify parallel lines, transversals, and corresponding angles.

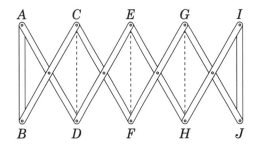

Alternate Teaching Approach
Give students sheets of tracing paper. Ask them to trace the angles formed by parallel lines and a transversal. Copy them on cardboard, and cut out models of different angles. They can place these models on top of other angles to find congruent angles.

Error Analysis
Students may mix up terms and definitions such as that for alternate interior angles. To help reduce confusion, encourage students to use manipulatives as well as a verbal and a visual approach when studying. They could use tracing paper or cardboard models or draw on the chalkboard.

Name _____ Date _____

Chapter 2   Test

Use the figure at the right to answer exercises 1 to 5.

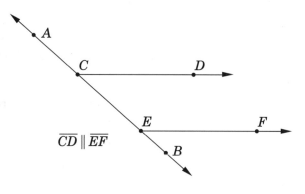

$\overline{CD} \parallel \overline{EF}$

1. Name a straight angle. _____
2. Name an acute angle. _____
3. Name an obtuse angle. _____
4. Name a pair of supplementary angles. _____
5. Name a pair of congruent angles. _____

In the figure at the right, $l \parallel m$ and $m\angle 8 = 135°$. Find the measure of each angle.

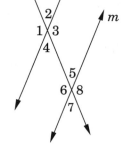

6. $m\angle 1 =$ _____   7. $m\angle 4 =$ _____

8. $m\angle 2 =$ _____   9. $m\angle 3 =$ _____

10. $m\angle 5 =$ _____   11. $m\angle 6 =$ _____

12. $m\angle 7 =$ _____

13. Draw the midpoint of $\overline{GK}$.
14. Draw the angle bisector of $\angle WXY$.

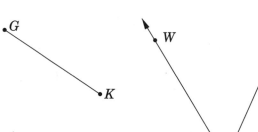

15. When a light beam hits a flat mirror, two congruent angles are formed by the path of the beam. The diagram at the right shows the path of the beam.

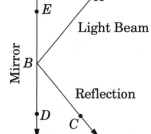

   a. Name all pairs of congruent angles shown.

   _____

   b. Why is this concept important?

   _____

   _____

Chapter 3 Triangles

Introduction

Challenge students to recreate the right triangle formed by the ancient Egyptians. Provide them with clay, three pegs or dowels, string, and a ruler. Review the meaning of the term right angle. *Explain that a right triangle has one right angle. After completing Lesson 3.6, return to the triangles students made. Ask students to explain how the rope helped the Egyptians know they had formed a right triangle.* [The square of the side opposite the right angle (5^2) equals the sum of the square of the sides that form the right angle ($3^2 + 4^2$).]

Lesson 3•1 Identifying the Line Segments That Form Triangles

Chalkboard Activity

To review line segments and angles, draw and label six points on the chalkboard. Challenge students to connect the points to draw as many line segments as possible. Then have them use the points to identify angles by naming the angles created by the line segments. You might also encourage students to identify shapes, such as squares and triangles, that are formed by the line segments they drew.

Using Manipulatives

Give each student three toothpicks. Ask them to use the toothpicks to make triangles. Then ask them to make a figure that is not a triangle. Encourage students to record their work, including sketches of the figures they make. [A non-triangle figure would be the three toothpicks placed in a straight line or with one open vertex.]

Alternate Teaching Approach

Ask students to construct triangles on a geoboard and record them on geodot paper. Use their triangles to identify sides, vertices, interior points, and exterior points.

Final Check

Ask students to identify the vertices and sides of $\triangle MNO$. [vertices: points *M*, *N*, and *O*; sides: $\overline{MN}$, $\overline{NO}$, $\overline{OM}$]

Lesson 3•2 Classifying Triangles

Chalkboard Activity

Read aloud the directions below. Allow students to show their work on the chalkboard.

1. Draw acute $\angle ABC$. Then draw line segment *AC*. Describe the triangle formed.

2. Draw obtuse $\angle RST$. Then draw line segment *RT*. Describe the triangle formed.

3. Draw right $\angle MNP$. Then draw line segment *MP*. Describe the triangle formed.

Using Manipulatives

Ask each student to make a right triangle on a geoboard. Then direct students to move one vertex of the triangle to make an acute triangle. Ask them to continue this process to make an obtuse triangle, a scalene triangle, and an isosceles triangle. Let students draw the triangles they make on geodot paper.

Working with ESL/LEP Students

Students can make a reference list of the different triangles discussed in this lesson. Encourage them to draw pictures and include symbols in their list.

Final Check

Students can draw and label the six kinds of triangles on separate index cards. Have students exchange cards and use a protractor and ruler to verify the identity of each triangle. Allow students time to discuss among themselves any discrepancies they may have discovered.

Lesson 3•3 Identifying Congruent Triangles

Chalkboard Activity

As a review of triangles and their parts, assign different kinds of triangles for students to draw on the chalkboard. Ask them to label the triangles' vertices. Then have other students identify the line segments and angles that make up each triangle.

Alternate Teaching Approach

Invite a group of students to the chalkboard to draw examples of the following kinds of triangles: equilateral, isosceles, scalene, right, acute, and obtuse. Then ask another group to draw matching triangles that have the same size and shape. Explain that the matching triangles are called congruent triangles. Ask students how they could verify that two triangles are congruent. [Measure the sizes of the angles and the lengths of the sides.]

Final Check

Give students a copy of a triangle named *MNP*. Ask them to draw a triangle congruent to △*MNP* and name it △*RST*. Then ask them to draw two congruent triangles of their own. Check their drawings to be sure students understand the concept of congruency.

Lesson 3•4 Proving Triangles Congruent

Chalkboard Activity

As a review of congruency, ask students to draw each of the following figures.

1. Congruent line segments: $\overline{DE}$ and $\overline{KL}$
2. Congruent angles: ∠*ABC* and ∠*GHI*
3. Congruent triangles: △*JKL* and △*XYZ*

Using Manipulatives

Students can work in small groups. Give each group several toothpicks. Direct students to place two toothpicks together to make two sides of a triangle. Then ask them to find how many other sides can be drawn to complete the triangle. [They should find there is only one other possible side.] Ask students to name the postulate they demonstrated in this activity. [SAS] Then ask if they can use the toothpicks to show any of the other postulates, and, if they can, to demonstrate it. [The SSS postulate can be demonstrated, but the ASA postulate cannot.]

Working with ESL/LEP Students

Students may mix up the order of letters in the abbreviations of the postulates. Point out that except for the SSS postulate, there is always one different letter between two that are the same (SAS and ASA). Emphasize that learning the letters in the proper order is a simple and effective way to remember the postulates. Encourage students to memorize the abbreviations for the postulates.

Error Analysis

When using the SAS postulate, students might evaluate the wrong angles. Emphasize that the A in the SAS postulate refers to the angle that is *between* the two congruent sides of a triangle.

Lesson 3•5 Angles of a Triangle

Chalkboard Activity

As a lead-in to the lesson, have students measure and label the angles of several triangles drawn on the chalkboard. Encourage them to add the measures of the angles of each triangle and make a general statement about what they discover. [The sum of the measures of the angles of a triangle is 180°.]

Using Manipulatives

Invite students to cut out paper triangles of various shapes and sizes. Direct them to fold each triangle into a rectangle, using the steps Quincy followed in the lesson. Discuss why Quincy could fold every triangular kite the same way, regardless of its shape or size. [The sum of the angles of every triangle is equal to 180°.]

Error Analysis

As students find the missing measure of an angle in a triangle, be sure they add the two known measures together *before* subtracting from 180°. To emphasize the importance of this procedure, ask them to evaluate the following expressions and compare the results of the two expressions in each exercise.

1. 180° − 65° + 30° [145°]
   180° − (65° + 30°) [85°]

2. 180° − 120° + 10° [70°]
   180° − (120° + 10°) [50°]

Lesson 3•6 Right Triangles

Chalkboard Activity

Before learning about the Pythagorean Theorem, students may benefit from a review of square roots. Ask them to find the square root of the following numbers.

1. $\sqrt{36}$ [6]   2. $\sqrt{49}$ [7]
3. $\sqrt{100}$ [10]   4. $\sqrt{81}$ [9]

Using a Calculator

The use of a calculator is suggested in exercises 11 to 16. Examples of the steps that can be used to complete these exercises with a calculator are shown below. Note: Some calculators have a 🔲 key; other calculators use 🔲🔲 to calculate the square root.

1. 4 🔲 🔲 8 🔲 🔲 🔲 🔲 [8.94427191]

Display after first 🔲 will be 80.

2. 10 🔲 🔲 5 🔲 🔲 🔲 🔲 [8.660254038]

Display after first 🔲 will be 75.

Error Analysis

When using the Pythagorean Theorem, students may forget to square the lengths of the sides of the triangle. For example, for a right triangle with legs

5 and 6, students might calculate the hypotenuse to be 11 instead of $\sqrt{61}$, or 7.8. Remind them that the Pythagorean Theorem refers to the *squares* of the sides of a right triangle.

Lesson 3•7 Similar Triangles

Chalkboard Activity

As an introduction to similar triangles, read aloud the following directions.

1. Draw two non-touching line segments of different lengths.

2. Using each line segment you drew as the base, construct an angle of measure 60°.

Invite students to construct each figure on the chalkboard. Ask them to describe how the figures are similar but not identical. [The angles have the same measure, but the sides have different lengths. In other words, the angles are congruent, but the sides are not.]

Using Manipulatives

Tangrams can be used to show the difference between congruent and similar triangles. As students explore triangles, make sure they recognize that the two small triangles are congruent, that the angles of the four triangles are congruent, and that the proportion of the small triangle to the medium triangle and large triangle is $\frac{1}{2}$ and $\frac{1}{4}$, respectively.

Using a Calculator

Students can use calculators to solve proportions. The steps below can be used to solve the proportion $\frac{2}{4} = \frac{x}{6}$.

2 ⊠ 6 ⊞ 4 ▣ [3]

Alternate Teaching Method

Have each student draw a triangle. If you have access to a copy machine, make one enlarged and one reduced copy of each triangle. Be sure to write the percent of enlargement or reduction on each copy. Encourage students to measure the angles and sides of the triangles. They should observe that the corresponding angles of the triangles are congruent, whereas the sides are not. Lead them to recognize that the corresponding sides are proportional.

Error Analysis

In exercises 13 to 16, watch that students set up proportions that compare *corresponding* sides of the triangles. For example, in exercise 13, students might compare side $\overline{EF}$ with side $\overline{WY}$,

rather than side $\overline{XY}$. To avoid this error, you might have them list the corresponding sides of each pair of triangles before doing the exercises.

Final Check

For the following similar triangles, ask students to complete the statement of similarity, identify the triangles' corresponding parts, and use proportions to find the missing lengths.

1. $\triangle ABC \sim [\triangle YZX]$

2. $\overline{AB} \leftrightarrow [\overline{YZ}]$
   $\overline{BC} \leftrightarrow [\overline{ZX}]$
   $\overline{CA} \leftrightarrow [\overline{XY}]$

3. $c = [9]$ $y = [8]$

Lesson 3•8 Using Inequalities for Sides and Angles of a Triangle

Chalkboard Activity

Invite students to work in pairs at the chalkboard. Direct one student in each group to draw and label a triangle. Then have the other student identify the angle that is opposite each side of the triangle. Allow students to repeat the activity, switching roles. You might wish to extend the activity by having students observe that the angle opposite the longer of two sides of a triangle is greater than the angle opposite the shorter of those two sides.

Working with ESL/LEP Students

Choose three points in the classroom to represent Marta's house, her friend's house, and her school as described in the lesson opener. Then ask students to act out the situation. Have them measure the distances to prove that the sum of any two parts of the trip is greater than the third part. Encourage students to describe the results in their own words.

Alternate Teaching Strategy

Students can work in small groups, with each member drawing a triangle. Direct the groups to measure each side of each triangle, find the sum of two of the sides, and compare the sum to the length of the third side of the triangle. Suggest that students repeat the activity, using the sums of two other sides of the triangle. Encourage groups to discuss their observations among themselves and write a statement describing what they discovered about triangles. [The sum of the lengths of any two sides of a triangles is greater than the length of the remaining side.]

Chapter 3   Test

Classify each triangle as *acute, obtuse, right, scalene, isosceles,* or *equilateral.* Use as many terms as apply to each triangle.

1.

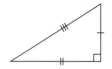

_____

2.

_____

3.

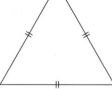

_____

4.

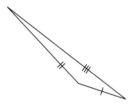

_____

5.

_____

6.

_____

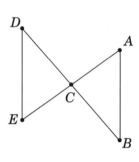

7. Name the congruent, corresponding parts needed to show that ΔECD and ΔACB are congruent according to the SSS postulate. _____

8. Use the Pythagorean Theorem to find the missing measure in the triangle at the right. Use a calculator to find the measure to the nearest tenth. _____

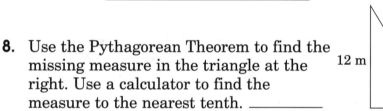

In ΔRST, *m∠2* = 50° and *m∠3* = 80°.
Find the missing measures.

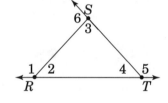

9. *m∠4* = _____

10. *m∠1* = _____

11. *m∠5* = _____

12. *m∠6* = _____

ΔABC is similar to ΔDEF.
Use proportions to find the missing sides.

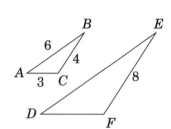

13. $\overline{DE}$ = _____

14. $\overline{DF}$ = _____

15. Can the numbers 1, 2, and 3 be the sides of a triangle? Why or why not?

_____

_____

Chapter 4 Polygons

Introduction

Point out that mosaics were used in ancient times. Mosaic floors from Roman times were found at Pompeii. Mosaic work was used for church decorations during the Middle Ages. During the Renaissance period they were used to imitate paintings. Ask students to look for mosaics in their area. They might try churches, universities, and art museums. They should note or draw the shapes used for the various mosaic tiles. Compare these polygons to the ones discussed throughout the chapter.

Lesson 4•1 Types of Polygons

Chalkboard Activity

Ask students to name as many geometric figures as they can. Encourage them to name polygons they are familiar with, such as squares, rectangles, pentagons, and hexagons. Write the responses on the board. Next to each, ask volunteers to write the number of sides the figure has.

Using Manipulatives

Invite pairs of students to draw different polygons. One student draws the polygon and gives the number of sides. The partner names the polygon. Then partners exchange roles and repeat the activity.

Working with ESL/LEP Students

Encourage students to draw and describe the polygons listed in the chart on page 90. If necessary, challenge them to define each polygon in their primary language on index cards. Students can refer to these cards when necessary.

Alternate Teaching Approach

Challenge students to find examples of polygons in the environment, magazines, or newspapers. Examples may include a bicycle frame for triangle, a field for quadrilateral, a school crossing sign for pentagon, and a stop sign for octagon. Invite students to share their findings.

Lesson 4•2 Properties of Parallelograms

Chalkboard Activity

To lead into this lesson, challenge students to construct the figures described on the board.

1. Draw quadrilateral $ABCD$.
2. Draw $\overleftrightarrow{MN} \parallel \overleftrightarrow{PQ}$.
3. Use the parallel lines you drew in part 2 to draw quadrilateral $MNQP$.

Ask students, *Is it possible to draw a quadrilateral with two pairs of parallel sides? Why or why not?* [Yes, opposite sides can be parallel.]

Working with ESL/LEP Students

Encourage students to make lists of the properties of parallelograms, using their own words. If necessary, students can include sketches and symbols in the descriptions. Emphasize that sketches and symbols will help them communicate their ideas.

Alternate Teaching Approach

Have students work in small groups to draw parallelograms on large sheets of paper. As the properties of parallelograms are introduced, students add them to their drawings, using appropriate symbols. Let students make copies of their drawings. Encourage them to cut out sections that demonstrate the properties of diagonals.

Error Analysis

As students are naming congruent line segments and angles, check to see if they are using the correct labels. They may need to review how to use labels when naming segments and angles.

Lesson 4•3 Rectangles

Chalkboard Activity

To introduce this lesson, invite students to construct the figures described below.

1. Draw right angle XYZ.
2. Use this right angle to draw quadrilateral $XYZW$.
3. Draw right angle XYZ again. Use it to draw parallelogram $XYZV$. Ask, *What do you notice about the parallelogram?* [The parallelogram has four right angles.]

Working with ESL/LEP Students

Before beginning this lesson, you may wish to discuss the *terms rectangle, diagonal, congruent, parallel, bisect, parallelogram,* and *intersect.* Encourage students to talk about these terms, using their own definitions. If necessary, invite students to come to the board to draw diagrams explaining the terms.

Error Analysis

Some students may have trouble understanding that all rectangles are parallelograms, but not all parallelograms are rectangles. Emphasize that a rectangle is a special type of parallelogram. Ask students to compare drawings of rectangles with drawings of parallelograms. Suggest they work in small groups to discuss how they are the same and how they are different. Encourage groups to make a two-column list, labeling one side *Similarities* and the other side *Differences*.

Lesson 4•4 Rhombuses and Squares

Chalkboard Activity

Before beginning this lesson, write the terms *parallelogram* and *rectangle* on the board. Invite students to define each term. Write their definitions on the board. [A parallelogram is a quadrilateral with two pairs of parallel sides. A rectangle is a parallelogram with four right angles.] After students define each term, challenge them to define square. [A rectangle with four congruent sides.] Write the definition of square on the board.

Using Manipulatives

Use four paper strips of equal length. Connect them with brads to form a rhombus. Give each small group a rhombus to manipulate. Ask students to form different rhombuses and record their observations. Have them discuss the following questions.

1. *Is a square a rhombus?*
   [yes]

2. *Are all squares rhombuses?*
   [yes]

3. *Is a rhombus a square?*
   [sometimes]

4. *Are all rhombuses squares?*
   [no]

5. *What is the difference between a square and a rhombus that is not a square?*
   [A rhombus that is not a square does not have four right angles.]

Error Analysis

Some students may not understand the differences between rhombuses and squares. Encourage them to define rhombus and square in their own words. They could draw pictures to help clarify their definitions. If students are still struggling with explaining the differences between rhombuses and squares, use the manipulative described above to help them see the differences.

Lesson 4•5 Trapezoids

Chalkboard Activity

Write the following exercises on the board and ask students to construct the figures described below.

1. Draw parallel lines *HJ* and *KL*.

2. Draw isosceles triangle *MNP*.

3. Draw line segment *DF* with midpoint *E*.

Using Manipulatives

Give small groups of students a geoboard and dot paper. Ask them to use rubber bands to make a trapezoid on the geoboard and then draw the trapezoid on dot paper. Challenge them to label the trapezoid on the dot paper, identifying the bases, legs, median, and base angles. Groups should note whether the trapezoid is isosceles and justify their answers. Repeat the procedure to make more trapezoids by moving the vertices along the parallel lines that contain the original bases.

Working with ESL/LEP Students

Encourage students to use their own words to define terms such as *trapezoid, base, base angle,* and *isosceles trapezoid* on note cards. Be sure they include drawings and symbols. Help them compare drawings of trapezoids with parallelograms, describing similarities and differences.

Error Analysis

For exercises 13 to 16, make sure students add the base numbers before they divide by 2 to find the measure of $\overline{LM}$. Point out that to find the measure of the bases, they can write an equation and solve for the missing variable.

Lesson 4•6 Regular Polygons

Chalkboard Activity

To review polygons, ask students to draw each figure.

1. Square

2. Pentagon

3. Hexagon

4. Octagon

Using Manipulatives

Invite groups of students to use a compass, a protractor, and a straightedge to construct regular polygons. Ask each group to devise its own way to do this construction. You may wish to suggest that group members begin with a circle and separate it into equal arcs corresponding to the number of sides of the polygon they want to construct. Remind students that a circle contains 360°. After students have completed several constructions, ask them to share their work with the class.

Working with ESL/LEP Students

Before beginning this lesson, ask each student to give a general definition of the term *regular*. Discuss how the definitions apply to polygons. After this discussion, help students define other terms in this lesson, such as *interior angles* and *exterior angles*. Encourage them to include drawings and symbols whenever possible.

Error Analysis

Make sure students use the correct formula for exercises 3 to 10. Point out that the directions are different. For exercises 3 to 6, the directions ask them to find the sum of the interior angles of each regular polygon. So, students would use the formula, sum = $180(n - 2)$. For exercises 7 to 10, the directions ask to find the measure of each interior angle. Therefore, students should use the formula, $\frac{180(n - 2)}{n}$.

Lesson 4•7   Congruent Polygons

Chalkboard Activity

Draw and label congruent $\triangle ABC$ and $\triangle DEF$ on the board. Invite students to complete the following exercises.

1. Name all corresponding sides.
   [$AB \cong DE$, $BC \cong EF$, $CA \cong FD$]

2. Name all corresponding angles.
   [$\angle A \cong \angle D$, $\angle B \cong \angle E$, $\angle C \cong \angle F$]

Alternate Teaching Approach

Give students several sheets of tracing paper. Ask them to trace and label a polygon. Then they can place it on top of another polygon to see if the polygons are congruent. If the polygons are congruent, students can use the corresponding labels on the polygons to name the corresponding parts.

Error Analysis

For exercises 1 to 10, some students may have trouble naming corresponding parts. Before they begin the exercises, encourage students to trace and label pentagon *PQRST* and place it on top of pentagon *FGHJK*. Then have them complete the exercises.

Lesson 4•8   Similar Polygons

Chalkboard Activity

As a preview for the lesson write the following proportions on the board. Invite students to cross multiply to solve for the given variables.

1. $\frac{x}{8} = \frac{15}{24}$
   [$x = 5$]

2. $\frac{7}{x} = \frac{35}{10}$
   [$x = 2$]

3. $\frac{2.5}{3} = \frac{x}{6}$
   [$x = 5$]

4. $\frac{2}{3} = \frac{5}{x}$
   [$x = 7.5$]

Alternate Teaching Approach

Give groups of students several paper strips and brads. Ask groups to construct one polygon, then another that is similar. Groups can exchange their original polygons, challenging others to make similar polygons.

Error Analysis

For exercises 7 to 9, some students may have trouble setting up a proportion to find the missing measure. Remind students that corresponding sides of two similar polygons are in proportion. Have students list the corresponding sides before they set up the proportion.

Lesson 4•9   Drawing Polygons on the Coordinate Plane

Using Manipulatives

Suggest that students work in small groups with geoboards. The geoboard represents a coordinate plane, with the bottom left corner representing the origin, the bottom row representing the x-axis, and the column along the left side representing the y-axis. Students use rubber bands to show the path from the origin to a given point. Each group records its work on a drawn coordinate plane.

Working with ESL/LEP Students

Before you begin this lesson, you may wish to discuss the terms *transformation, translation* (slide), *reflection* (flip), and *rotation* (turn). Encourage students to describe these terms in their own words. Invite students to draw on the board a diagram of each term. Suggest that they define these terms on index cards, with a small diagram next to each definition.

Error Analysis

For exercises 10 and 11, some students may have trouble graphing the image of ΔFGH under the transformation described. Encourage students to trace ΔFGH and use the traced figure to help them with the exercises.

Final Check

Invite students to work in small groups. Have each group draw and label several points on a coordinate plane. Then have the groups exchange papers. Ask students to find the coordinates for the points and to name the polygons formed by the points.

Lesson 4•10 Tessellations of Regular Polygons

Using Manipulatives

Use equilateral triangles, squares, regular hexagons, and regular octagons to demonstrate tessellations. Shapes can be cut out of paper or cardboard. Give students time to experiment with these figures in small groups. Ask them to complete the following activities and answer the questions.

1. Use the triangles and squares to make a pattern. How many different patterns can you make?

2. Use the squares and octagons to make a pattern. Are there any gaps in your pattern?

3. Use the hexagons and triangles to make a pattern. Are there any gaps in your pattern?

4. What combinations of the figures above will make a gap?

Alternate Teaching Approach

Challenge students to make their own tessellations using the following procedure. Cut a 4 cm by 4 cm square. Cut a piece from the bottom and slide it directly up to the top. Tape that piece to the top of the square. Students should be able to tessellate their own shape by performing repeated slides.

Final Check

To be sure students understand why some figures tessellate and others do not, have students make a chart to explain which polygons tessellate, and which polygons together tessellate. Use the polygons mentioned on the bottom of page 90.

Name _____ Date _____

Chapter 4   Test

Match the name of each polygon with the best figure. Use each letter only once.

1. quadrilateral _____
2. trapezoid _____
3. hexagon _____
4. rectangle _____
5. rhombus _____
6. octagon _____

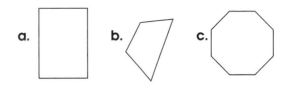

7. The measure of one angle of a parallelogram is 65°. What are the measures of the other angles of the parallelogram?

_____

8. The length of a base of a trapezoid is 23 km. The length of its median is 29 km. What is the length of the other base?

_____

The sum of the measures of the angles of a regular polygon is 540°.

9. How many sides does the polygon have? _____

10. What is the name of the polygon? _____

11. What is the measure of each interior angle of the polygon?

_____

12. What is the measure of each exterior angle of the polygon?

_____

13. If three congruent equilateral triangles are joined so that each shares at least one side, what polygon is formed?

_____

Use the similar quadrilaterals at the right to answer exercises 14 and 15.

14. $\overleftrightarrow{VT} =$ _____

15. $\overleftrightarrow{XT} =$ _____

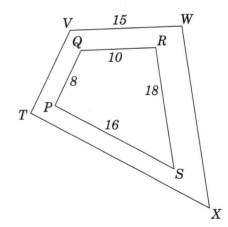

Chapter 5 Area and Perimeter of Polygons

Introduction

Draw the shape of a basketball court on the board, or use the drawing on page 126. Ask students to point out the different polygons and other shapes that make up the court. Elicit examples of other playing surfaces that include polygons. Students might draw these areas on the board. Invite them to take turns outlining some of the polygons included in the playing areas. [Examples include tennis courts; football and soccer fields; and a baseball diamond, including batter's box.]

Lesson 5•1 Perimeter of Parallelograms and Rectangles

Chalkboard Activity

List the following words on the board. Invite students to show what each term means using either a drawing or a written definition. Check using students' work.

parallelogram   rectangle   perimeter

length   width   diagonal

Using Manipulatives

Construct a worksheet containing several rectangles or parallelograms drawn on grid paper.

Provide students with pieces of string or yarn and the worksheet. Have them place a piece of string or yarn around each figure and then measure the length of the string. Next have students find the perimeter of each figure by measuring the lengths of the sides and using the rule to find each perimeter. Conclude by asking students to compare the results of each method for finding a perimeter.

Error Analysis

Some students may forget to add both lengths and both widths when finding the perimeter of a parallelogram or rectangle. Encourage students to label all sides of their drawings before calculating the perimeter. Some students may prefer to find the perimeter, using the rule $2l + 2w$ rather than $l + l + w + w$, in order to avoid this error.

Lesson 5•2 Area of Parallelograms and Rectangles

Using Manipulatives

Materials needed are pencils, grid paper, rulers, scissors. Invite students to draw a rectangle and a parallelogram on a sheet of grid paper. Ask them how they could divide each figure in half. If it is not already stated by a student, point out that one method is to draw a diagonal. Ask students what figures result when they divide along the diagonal. Invite students to cut out their rectangle and the parallelogram, cut along a diagonal, and put one area over the other to see if the areas are the same. [If the figure is divided in two along a diagonal, then the result is two pairs of congruent triangles.]

Working with ESL/LEP Students

Ask students to think about when they have heard the terms height and base used in everyday life. Ask questions such as:

What is your *height?*

Where is the *base* of the Statue of Liberty? [Show students a photograph of the statue.]

After a brief discussion of these terms, show students how they can apply these ideas to parallelograms.

Error Analysis

Be sure students understand the difference between the methods for finding the areas of rectangles and parallelograms. If given the length, width, and height of a parallelogram, some students may try to find the area by multiplying the length times the width. Be sure students understand that the length of a parallelogram is not the same as its height.

Lesson 5•3 Estimating Perimeter and Area

Chalkboard Activity

Review various kinds of polygons. Draw a triangle, a rectangle, a square, a pentagon, and a hexagon on the board and have students identify each figure. Next, challenge students to draw and label the dimensions of as many rectangles as they can with a perimeter of 40, using whole number dimensions. On the board, draw two rectangles to start them off, 5×15 and 1×19. Volunteers can be called to the board to draw and label other possible choices.

[2 × 18, 3 × 17, 4 × 16, 6 × 14, 7 × 13, 8 × 12, 9 × 11, 10 × 10]

Working with ESL/LEP Students

If your classroom floor has tiles, invite students to estimate the number of tiles on the floor. Or, if the hallway is tiled, pick two fixed positions and have students estimate the number of tiles on the hallway floor between these two locations. Challenge students to think of other ways to obtain more accurate values for these measurements. Compare these other ways with the students' methods of estimation to clarify what an estimate is.

Error Analysis

Students may not understand how to handle partial squares when estimating area and perimeter. Provide additional practice for students having problems with this estimation skill. Students can draw their own figures on grid paper, estimate the area, and then cut the figures apart, arranging the partial squares so that they have complete squares. They can compare the area of their cut-up figures with their estimates.

Final Check

Ask each student to draw a polygon on grid paper. Then have students exchange and estimate the perimeter and area of each others' polygons. Invite volunteers to explain how they arrived at their answers.

Lesson 5•4 Perimeter and Area of a Rhombus or a Square

Chalkboard Activity

Draw a rhombus and a square on the board. Have students volunteer to describe similarities and differences between the two figures. [Similarities students should observe are: both figures have 4 equal sides, 4 angles. Differences are: the square has 4 right angles; the rhombus does not.]

Working with ESL/LEP Students

To help students become more comfortable with the terms and ideas in this chapter, you might pair English-speaking students with non-English-speaking students to organize their understanding of the different quadrilaterals. Each pair can make a chart, using pictures and shapes to show the relationships among certain quadrilaterals. [Possible relationships: all squares are rectangles; all squares are rhombuses; all squares are parallelograms; some rhombuses are squares]

Students could share their work with the class. This activity should help everyone get a clearer understanding of the interrelationships among quadrilaterals.

Lesson 5•5 Properties of the Diagonals of a Rhombus

Chalkboard Activity

Draw the following figures on the board with dimensions labeled and have students find the area of each:

1. Rectangle with length 6.5 in. and width 8.2 in.
   [53.3in.²]

2. Rhombus with base 12 cm and height 4 cm
   [48 cm²]

3. Parallelogram with length 8 in.; base 10 in.; and height 6 in.
   [60 in.²]

Using a Calculator

Demonstrate that there are several different ways to use a calculator to evaluate the formula given in this lesson. Three different ways to use a calculator to find the area of the party invitations described in the opener are given below.

0.5 **x** 5 **x** 4.5 **=** 11.25

5 **x** 4.5 **÷** 2 **=** 11.25

1 **÷** 2 **x** 5 **x** 4.5 **=** 11.25

You may want to discuss with students why each of these methods produces the same result.

Error Analysis

Some students may find half of the length of each diagonal and then multiply. Review the order of operations to clarify the way the numbers should be multiplied. It may also be helpful to put parentheses around d_1 and d_2 in the formula.

Lesson 5•6 Area of Triangles

Alternate Teaching Approach

Invite students to work in pairs to complete the following:

1. Draw a parallelogram with one diagonal.

2. Measure the base and the height of the parallelogram.

3. Find the area of the parallelogram.

Then allow partners to discuss and answer the following questions:

1. What figures do you see on either side of the diagonal? [triangles]

2. How does the area of each of these figures compare to the area of the parallelogram? [The area of each figure is one-half the area of the parallelogram.]

3. Based on your answers to 1 and 2, write a rule to find the area of a triangle. Explain how you created your rule. [Students should realize that since the area of a triangle is $\frac{1}{2}$ the area of a parallelogram, the rule would be $\frac{1}{2} \times$ base $\times$ height.]

Working with ESL/LEP Students

Have students draw a triangle with the following terms labeled and shown: vertex, base, height, perpendicular, opposite base

Allow them to refer to their drawings as they work through this lesson.

Error Analysis

Some students may confuse the length of the sides with the height of some types of triangles. Be sure students use the measurement of the correct parts of a triangle when finding the area. To be sure students can properly identify the base and height of many different types of triangles, suggest they draw several triangles and label the base and height of each.

Lesson 5•7   Area of Trapezoids

Using Manipulatives

Have students draw five different trapezoids on grid paper. Next, ask them to draw and label at least two different line segments to represent the height of each trapezoid. Then ask students to estimate the area of each trapezoid they drew. As an extension, have them find the areas, using the rule $\frac{1}{2}h(b_1 + b_2)$. Finally, have students compare their estimates with the answers they found, using the rule.

Lesson 5•8   Perimeter and Area of Irregular Figures

Chalkboard Activity

Draw and label the polygons on the board. Invite students to tell how they would find the area of

each. Then have them find the areas.

Parallelogram: length = 6 cm; width = 8 cm; height = 4 cm
[Multiply length $\times$ height; Area = 32 cm^2]

Trapezoid: base 1 = 4 in.; base 2 = 6 in.; height = 8.5 in.
[$\frac{1}{2}h(b_1 + b_2)$; Area = 42.50 in.^{2}]

Rhombus: diagonal 1 = 6.5 cm; diagonal 2 = 7 cm
[$\frac{1}{2}(d_1d_2)$; Area = 22.75 cm^2]

Alternate Teaching Approach

Students can work in pairs. Have each pair draw two different polygons on grid paper. Then, cut out the polygons, arrange them together, and trace their combined outline onto another sheet of grid paper. Next, find the area of each polygon by either counting the grids or using a rule for area. Finally, ask students to discuss and make a conclusion about the areas of the two polygons and the area of the figure you traced. [Students should conclude that the irregular figure's area is equal to the area of polygon 1 + area of polygon 2.]

Lesson 5•9   Determining the Difference Between Perimeter and Area

Alternate Teaching Approach

Invite students to work in pairs. Have one partner describe a situation in which the perimeter of a figure must be found. Have the other partner make up a situation in which the area must be found. Then have partners work together to write out each situation as a word problem, including dimensions. Finally, have pairs of students exchange and solve each others' problems.

Final Check

Ask questions similar to the following.

1. If we wanted to put a strip of colored tape around the edge of the classroom, would we need to find the perimeter or the area? [perimeter]

2. If we wanted to cover the floor with wall-to-wall carpeting, what would we need to know? [area]

Chapter 5 Test

Name _____ **Date** _____

Find the perimeter and area of each figure.

1.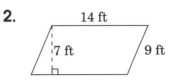

25 in.

$P =$ _____

$A =$ _____

2.

14 ft

7 ft   9 ft

$P =$ _____

$A =$ _____

3.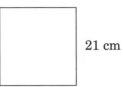

21 cm

$P =$ _____

$A =$ _____

4.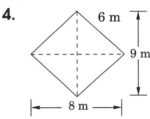

6 m

9 m

8 m

$P =$ _____

$A =$ _____

Find the area of each figure shown.

5.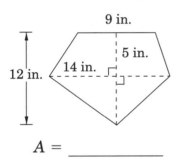

9 in.

5 in.

12 in.   14 in.

$A =$ _____

6.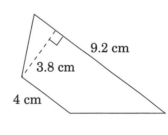

9.2 cm

3.8 cm

4 cm

$A =$ _____

Meg wants to wallpaper her bedroom. The room is 12 feet long and 15 feet wide with a ceiling that is 8 feet high. The wallpaper she wants comes in double rolls with 200 square feet of wallpaper in each roll.

7. How many square feet of wallpaper will Meg need to cover all four walls in her room? _____

8. Will Meg have enough wallpaper to cover her walls if she buys two double rolls of wallpaper? Why or why not? _____

9. If Meg puts a border around the entire room, how many feet of border paper will she need? _____

10. A floor that is 10 ft by 16 ft will be covered with triangular shaped carpet pieces shown in the diagram at right. How many pieces of triangular shaped carpeting will be needed to cover the floor? _____

16 ft

2 ft
1.5 ft   *Floor*   10 ft

Chapter 6 Circles

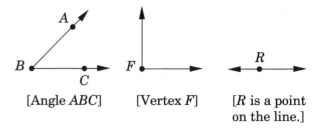

[Angle *ABC*] [Vertex *F*] [*R* is a point on the line.]

Introduction

Invite students to study the picture of the Aztec calendar. Display a string 3.6 meters long to give students an idea of the size of the calendar. Then ask them to draw a line on page 151 through the center of the calendar that divides it into two identical halves. You might mention that a line that divides a circle in half is called a diameter. In each half, students should be able to identify ten symbols that represent ten of the twenty days of the Aztec month.

Lesson 6•1   Parts of a Circle

Chalkboard Activity

As a lead-in to the lesson, draw the following on the board and have students identify each, using the letters given.

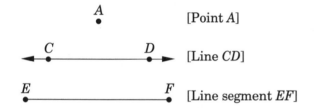

[Point *A*]

[Line *CD*]

[Line segment *EF*]

Working with ESL/LEP Students

Encourage students to explain each new term presented in this lesson in their own words. Ask them to discuss ways they can remember the meanings of these terms. You might have students make labeled drawings to help them remember the terms. Allow students to refer to their drawings throughout the chapter, as necessary.

Error Analysis

Some students may not remember that in order for a line to be a diameter of a circle, it must pass through the center of the circle. Present several circles with lines that are diameters and some with lines that are not. Have students identify which circles have lines that represent their diameters.

Lesson 6•2   More Parts of a Circle

Chalkboard Activity

Draw the following. Have students identify what the letters in each drawing represent.

Using Manipulatives

Provide small groups of students with copies of several different large circles, tape, scissors, and pieces of yarn that they can use to model the parts of a circle. For their finished product, each group should have a labeled model or models showing at least one chord, arc, central angle, tangent, and secant.

Working with ESL/LEP Students

Review the meanings of the terms introduced in this lesson, using the circle on page 156 as a reference. Encourage students to add examples of these terms to the labeled drawing they made in Lesson 6•1.

Error Analysis

Some students may have difficulty keeping track of the various angles and arcs in exercises 13 to 19. Encourage them to copy circle *X* and label the given information before they begin the exercises. You may also need to remind them that a circle is 360°.

Final Check

Discuss the following questions to assess students' understanding of the lesson concepts.

1. What is the difference between a chord and a secant? [A chord is a line segment with its endpoints on a circle. A secant is a line with two of its points on a circle.]

2. What is the difference between a tangent and a secant? [A tangent is a line that intersects a circle at one point. A secant is a line that intersects a circle at two points.]

3. How does increasing the measure of a central angle affect the minor arc of the angle? [Increasing the central angle increases the minor arc to the same degree.]

Lesson 6•3   Circumference of a Circle

Chalkboard Activity

Draw a circle, and show a diameter and a radius. Measure each and ask students to summarize the relationship between the diameter and radius. [The length of the diameter of a circle is twice the length of its radius.]

Using Manipulatives

Have students work in pairs to draw some circles of different sizes, either by tracing around various objects (quarter, base of a soda can, wristwatch, etc.) or by using a compass. Ask students to find the circumference of each circle by using a tape measure around the circle, and then by measuring the diameter and multiplying by π. Have students compare their results from using both methods.

Using a Calculator

Some calculators have a π key. Allow students to experiment with this key and describe the results. [They will find that the value for π has more than two decimal places.] Ask students to predict what will happen if they use the π key to find the circumference of a circle, rather than using the numeral 3.14. After they answer this question, invite students to experiment to see if their prediction is correct.

Error Analysis

Students often forget to multiply the radius by 2 when finding circumference. To help students avoid this error, suggest they write the formula they will use for each exercise before they begin their calculations.

Final Check

Ask students to estimate the circumference of each of the following circles.

1. Circle A with a diameter of 21 in. [About 64 in.]

2. Circle B with a radius of 1.5 cm [About 10 cm]

3. Circle C with a radius of 30 in. [About 200 in.]

Lesson 6•4   Area of a Circle

Chalkboard Activity

Review the formulas for finding the areas of a rectangle, square, and triangle. Then draw a figure composed of a 2 ft by 3 ft rectangle on which is set a triangle of base 2 ft and height 2 ft. The figure should look like a rocket. Challenge students to explain how to calculate the area of the figure and tell which formulas to use. [Separate the figure into a 2 ft by 3 ft rectangle and a triangle of base 2 ft and height 2 ft. Find the total area by using A rectangle $= l \times w = 2 \times 3 = 6$ ft^2 and A triangle $= \frac{1}{2} b \times h = \frac{1}{2} \times 2 \times 2 = 2$ ft^2. Total area of the figure is 8 ft^2.]

Error Analysis

Students may confuse the formulas for finding the area and circumference of a circle. To avoid this error, suggest that they write the formula for area as $\pi \times r \times r$ and then substitute the value for each term. Another common error is using the diameter instead of the radius in the area formula. Writing out the formula might help to remind students to use the value of the radius.

Lesson 6•5   Circumscribed and Inscribed Polygons

Chalkboard Activity

Draw one polygon that is circumscribed about a circle and another polygon of the same type that is inscribed in the circle. Ask students to describe similarities and differences between the circumscribed and inscribed polygons. [Inscribed polygons have their vertices on the circle. Circumscribed polygons have their sides tangent to the circle. Circumscribed polygons are larger than inscribed polygons for the same circle.]

Working with ESL/LEP Students

Have these students research the meanings of the prefixes *in-* and *circum-* and the word *scribe*. Ask them to share their findings with the class. Encourage students to combine these meanings to come up with definitions for the words *inscribed* and *circumscribed*.

Lesson 6•6   Measurement of Angles Circumscribed and Inscribed in a Circle

Chalkboard Activity

Draw the figure below. Challenge students to identify all the inscribed and circumscribed polygons. [Circumscribed: square, triangle; inscribed: square, two triangles]

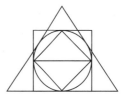

Final Check

Ask each student to write one exercise involving a circumscribed angle and one involving an inscribed angle. Have students exchange and solve each other's exercises.

Chapter 6 Test

Draw each figure described below.

1. $\odot A$ with radius $\overline{AB}$ and diameter $\overline{CD}$

2. $\odot N$ with tangent $\overleftrightarrow{OP}$ and chord $\overline{AB}$

3. $\odot O$ with central angle $\angle AOB$ and secant $\overleftrightarrow{AB}$

4. $\odot W$ with inscribed angle $\angle BOD$ and secant $\overleftrightarrow{OD}$

In this figure $m\overset{\frown}{AC} = 140°$. $\overleftrightarrow{AB}$ and $\overleftrightarrow{BC}$ are tangent to the circle and $m\overset{\frown}{DC} = 110°$. Find each measure.

5. $m\overset{\frown}{ADC} =$ _____

6. $m\angle ABC =$ _____

7. $m\overset{\frown}{DAC} =$ _____

8. An angle formed by two tangents to $\odot P$ intercepts arcs of measures 80° and 280°. What is the measure of the angle? _____

9. A circle has a radius of 12 centimeters. Find the circumference and area of the circle. _____

10. Mel plans to buy circular tiles for his patio. Tile A has a diameter of 10 inches. Tile B has an area of about 82 square inches. Which tile will cover a greater area? Explain your answer.

_____

_____

_____

Chapter 7　Recognizing Three-Dimensional Shapes

Introduction

Review with students the characteristics of a polygon. Explain that polygons have three or more straight sides and contain three or more angles. A polyhedron is a three-dimensional shape formed by assembling polygons. Introduce students to the five polyhedra on page 177. Identify each polyhedron and count the number of faces. [tetrahedron, 4 faces; octahedron, 8 faces; hexahedron, 6 faces; icosahedron, 20 faces; dodecahedron, 12 faces]

Lesson 7•1　Properties of Polyhedra

Chalkboard Activity

Draw a square on the board. Encourage students to discuss the characteristics of a square polygon. Extend the diagram to make a cube. Invite students to list its characteristics, including the number of edges, vertices, and faces. Draw a triangle and then a prism and discuss how their characteristics change.

Working with ESL/LEP Students

Because there are many terms to learn throughout this lesson, students may need more time to complete the activities. They may benefit from labeled models or drawings showing the terms listed in Words to Learn. Students can refer to these models as they work through the lesson. Encourage students to draw pictures of polyhedra and label the parts.

Alternate Teaching Approach

Exhibit examples of polyhedra such as boxes and other containers. Encourage students to bring in some examples of their own. On index cards, label each example by the type of polyhedron it represents. List its identifying characteristics, such as number of faces, edges, and vertices. Nearby, display objects that are not polyhedra. Use index cards to indicate why each object is not a polyhedron.

Error Analysis

Watch for students who make the assumption that any three-dimensional shape is a polyhedron.

For exercise 1, guide students to see why the first and third shapes cannot be considered polyhedra, although they are three-dimensional. [All of their flat surfaces are not polygons.]

Final Check

Show students a three-dimensional drawing of a prism with the vertices labeled with letters. Invite students to name the bases, faces, edges, and an altitude of the prism. Repeat with other drawings of polyhedra. Include a drawing of a three-dimensional shape that is not a polyhedron to see if students recognize the difference.

Lesson 7•2　Cubes and Rectangular Prisms

Chalkboard Activity

Draw a square and rectangle on the chalkboard; then discuss their similarities and differences. [opposite sides are parallel and equal, 4 right angles; rectangles have two pairs of congruent sides whereas squares have all 4 sides congruent]

Using Manipulatives

Provide students with several nets of cubes and rectangular prisms. Have them fold some of the nets and describe each prism made. Challenge students to predict what the prisms formed by the remaining nets will look like. They should fold the nets to check their predictions. As an extension, students could draw their own nets.

Final Check

List on the chalkboard the specifications of a cube or a rectangular prism without drawing its shape. Include the number of faces and their measurements. Invite students to guess what type of prism it is. Then ask a volunteer to draw it on the chalkboard or hold up a box matching the specifications listed.

Lesson 7•3　Pyramids

Chalkboard Activity

Draw a regular pyramid on the chalkboard and label its vertex and the angles around its base. Introduce the Words to Learn on page 184 by defining them with the help of the diagram. Emphasize the difference between a pyramid and a regular pyramid. [They both have a base and triangular lateral faces, but the regular pyramid has a regular polygon as its base and congruent isosceles triangles as its lateral faces.]

Alternate Teaching Approach

Provide students with several nets of different pyramids. Have them fold some of the nets and describe each pyramid. Encourage students to predict what the base of the pyramids formed by the remaining nets will look like.

Error Analysis

Watch for students who confuse the altitude with the slant height of a pyramid. Use the figure shown in exercises 1 to 5 to explain the difference. Encourage students to measure each dimension so they can see the difference for themselves. Provide several examples of pyramids with both of these parts shown and invite them to name the altitude and the slant height in each example.

Final Check

Encourage students to draw several pyramids of their own. After labeling the parts with letters, invite them to exchange pyramids and name the base, lateral faces, altitude, slant height, and whether it is a regular pyramid.

Lesson 7•4   Cylinders and Cones

Chalkboard Activity

Draw a picture of a rectangular prism, a cube, a pyramid with a regular base, and a pyramid with a triangular base. Challenge students to name the figures. Then draw a cylinder and a cone. Discuss the characteristics of the five figures. Explain that the cylinder and cone are also solids, but they contain circles for bases. Label the parts of the cylinder and cone from the Words to Learn list on page 186. Discuss the meanings of these terms.

Using Manipulatives

Provide students with several nets of cylinders and cones. Have them fold some of the nets and describe each figure made. Encourage students to predict what the figures formed by the remaining nets will look like. They can fold the nets to check their predictions.

Working with ESL/LEP Students

Make models of a cylinder and cone that can come apart. Students can handle the models and point to their parts, including faces, circles, corners, altitude, and slant height. Invite students to bring in their own examples of cylinders and cones or to draw their own. They could then make a chart similar to the following to show the differences and similarities.

Parts	Cylinder	Cone
1 circular base		✔
2 circular bases	✔	
altitude	✔	✔
slant height		✔

Final Check

Invite students to draw several cylinders and cones of their own or bring in examples. Students can trade their examples and name the parts. Challenge them to measure the altitude and slant height.

Lesson 7•5   Spheres

Chalkboard Activity

Review the parts of a circle: center, radius, and diameter. Draw a circle on the chalkboard and ask students to label the parts. Challenge them to draw their own circles, and to include center M, radius $\overline{MN}$, and diameter $\overline{AB}$.

Using Manipulatives

Cut a spherical object in half (ball, orange, nerf ball, etc.). Organize students into pairs or groups of three and give each group a sphere. Encourage them to label the center, radius, and diameter. Students could then measure the radius and diameter of their spheres.

Error Analysis

Watch for students who do not understand the relationship between the radius and diameter of a sphere. Review the relationship between the radius and diameter of a circle. Use the exercises in the Chalkboard Activity to help students think about this relationship. Extend the concept to spheres.

Chapter 7   Test

Write the letter of the figure.

a.   b. 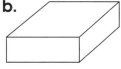   c.   d. 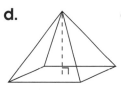   e.

1. cylinder _____   **2.** sphere _____

3. cone _____   **4.** rectangular prism _____

5. pyramid _____

Label each figure.

6. A cone with base circle P, altitude $\overline{MP}$, and slant height $\overline{MK}$.

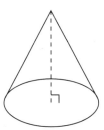

7. A cylinder with altitude $\overline{XY}$ and bases circle X and circle Y.

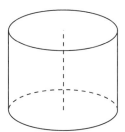

8. A regular pyramid with a rectangular base, altitude $\overline{AB}$, and slant height $\overline{AC}$.

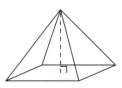

9. A sphere with center M, radius $\overline{MP}$, and diameter $\overline{OR}$.

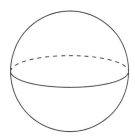

10. If you folded the outline of the shape shown along the dotted lines and glued the edges together, what shape would you have?

_____

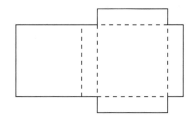

Chapter 8 Surface Area of Three-Dimensional Shapes

Introduction

Ask students to imagine gift wrapping a box. How would they determine how much wrapping paper to use? (Most students will estimate how much they would need by imagining how far the paper would fold over to cover the box.) If they underestimated, would it matter? (Probably not; they would start again or make do with a partly covered box.) Would it matter if they over-estimated? (No; they would have more paper that overlaps.) Now, what if students have a business that includes a gift wrapping service? Wasted paper becomes an expense. How would they minimize waste? (Students might suggest measuring the area of the different size boxes to get a more accurate estimate of how much paper to use for each box.)

Lesson 8•1 Surface Area of Rectangular Prisms

Chalkboard Activity

As a review write the following exercises on the board. Challenge students to find the area of each rectangle.

1. $l = 6$ ft, $w = 4$ ft $A =$ [24 ft^2]

2. $l = 7$ in., $w = 3$ in. $A =$ [21 in.^{2}]

3. $l = 3$ in., $w = 9$ in. $A =$ [27 in.^{2}]

4. $l = 4$ ft, $w = 10$ ft $A =$ [40 ft^2]

Using a Calculator

Students can use calculators to find the surface area. The method shown below can be used to find the surface area of Cara's box in the first lesson example.

2 [X] 3 [X] 2 [+] 2 [X] 6 [X] 2 [+] 2 [X]
6 [X] 3 [=] 72

Notice that the parentheses used in the example are not necessary. Encourage students to experiment with different methods.

Alternate Teaching Approach

Provide boxes or have students bring in boxes shaped like rectangular prisms or cubes. Suggest students work in small groups and measure the edges of several of the boxes. Then they can dis-cuss how the dimensions of the edges might be related to the surface area of each box. You may want to have students cut at least one of the boxes apart at the edges to form a net for the box. Emphasize that the area of this flat piece of card-board is the surface area of the box. You may want to save the boxes to use in Lesson 9•1.

Error Analysis

Some students may need a formal rule to help them remember how to find the surface area of prisms and cubes. Encourage them to create their own rule for each. [Rules should be in the following forms: Surface Area of Prism = 2 (area of A) + 2 (area of B) + 2 (area of C); Surface Area of Cube = 6s^2.]

Lesson 8•2 Lateral and Surface Area of Cylinders

Chalkboard Activity

As a review of circles, write the following on the board. Ask students to find the circumference of each circle.

Use 3.14 for π in the formula $C = \pi \times d$.

1. $d = 4$ in. $C =$ [12.56 in.]

2. $d = 8$ ft $C =$ [25.12 ft]

3. $d = 10$ in. $C =$ [31.4 in.]

4. $d = 100$ in. $C =$ [314 in.]

Using a Calculator

Students can use calculators to find the lateral and surface areas. The two methods shown below can be used to find the lateral area of the jar in the first lesson example.

3.14 [X] 10 [X] 12 [=] 376.8
[π] [X] 10 [X] 12 [=] 376.9911184

Discuss with students the accuracy of these two answers.

Alternate Teaching Approach

Provide containers shaped like cylinders or have students bring in containers such as empty cans. Have students work in small groups and measure the height and diameters of several containers. Then they can find the lateral area and surface area of each container. If you have a thin cardboard container, you may want students to cut it apart at the edges to form a net for the container. Emphasize that the area of the rectangular piece of cardboard is the same as the lateral area of the container and that the area of all of the cardboard is the surface area of the cylinder. You may want to save the containers to use in Lesson 9•2.

Lesson 8•3 Surface Area of Rectangular Pyramids

Chalkboard Activity

Write the following exercises on the board. Ask students to find the area of each triangle. Remind them that the area for a triangle is $\frac{1}{2}bh$.

1. $b = 5$ ft, $h = 6$ ft $A =$ [15 ft²]
2. $b = 4$ in., $h = 8$ in. $A =$ [16 in.²]
3. $b = 2$ in., $h = 10$ in. $A =$ [10 in.²]
4. $b = 8$ ft, $h = 9$ ft $A =$ [36 ft²]
5. $b = 12$ ft, $h = 5$ ft $A =$ [30 ft²]

Alternate Teaching Approach

Students can work in pairs or small groups to draw nets for one square pyramid and one rectangular pyramid on pieces of cardboard. Ask students to find the dimensions of the base of the pyramid and the height of each triangle and label them directly on the cardboard. Point out that the heights of the triangles are the slant heights of the pyramid. Then have students fold the nets to make the pyramids. Invite them to find the surface area of each pyramid, based on the dimensions they labeled.

Error Analysis

Because there are so many dimensions to keep track of, students may get confused about which to use. To help them remember which dimensions are needed to find lateral and surface areas of rectangular and square pyramids, invite students to draw a model of each figure, label each part, and write the rule for lateral and surface area below each model. Allow them to use models as reference tools as they work through this lesson.

Lesson 8•4 Lateral and Surface Area of Cones

Working with ESL/LEP Students

Before beginning this lesson you may wish to review several terms with students to be sure they understand the concepts taught up to this point. Talk about the following concepts, encouraging students to use the geometric terms in complete sentences for each answer. Use nets as visual references.

1. What is the difference between the lateral area of a cylinder and the surface area of a cylinder? [The lateral area does not include the top or bottom faces of the cylinder. The surface area is the total area of all faces of the cylinder.]

2. What is the net of a shape? [The net is the pattern of a shape that, when folded, will become a three-dimensional figure.]

3. What is the shape of the net of a cylinder? [a rectangle]

4. What shapes make up the net of a square pyramid? [four triangles and a square]

5. What shapes make up the net of a rectangular pyramid? [four triangles and a rectangle]

Error Analysis

Be sure students understand that if they are given the diameter of a cone they must first find the radius. Then they follow the rules given in this lesson to find the lateral and surface areas.

Final Check

Using exercise 19 about the cost of making funnels, challenge pairs of students to find the cost for each cone-shape product, such as party hats, ice-cream cone, paper cones for coffee makers, and water dispensers. Students should decide the size of the product and the cost per square inch for making that product. Invite them to share their ideas with other pairs of students in the class.

Lesson 8•5 Surface Area of Spheres

Chalkboard Activity

To review finding the area of circles, write the following exercises on the board. Invite students to find the area for each circle.

Use 3.14 for π in the formula $A = \pi r^2$.

1. $r = 4$ cm $A =$ [$A = 50.24$ cm²]
2. $r = 3.5$ cm $A =$ [$A = 38.465$ cm²]
3. $r = 10$ in. $A =$ [$A = 314$ in.²]

Error Analysis

Watch for students who use the diameter instead of the radius when finding the surface area. To avoid this error, have students complete several exercises in which you give them a diameter and they need to find the radius. Then have them find the surface of a sphere with each diameter. Repeat this process until students feel comfortable with the concept.

Final Check

Using exercise 8, challenge groups of students to find the surface areas of the other eight planets. Then have students compare the surface areas. Remind them that if the diameter is given, they need to find the radius to calculate surface areas. Encourage groups to share their findings.

Chapter 8   Test

Find the surface area and lateral area of each figure described. Show your work. Round answers to the nearest hundredth.

1. cylinder with diameter
   8.5 in. and height 9.5 in.
   $LA = \pi r h$
   $LA = $ _____
   $SA = LA + 2\pi r^2$
   $SA = $ _____

2. cube with edge of 4 cm
   $SA = 6s^2$
   SA _____

3. cone with diameter 14 cm
   and slant height 20 cm
   $LA = \pi r s$
   $LA = $ _____
   $SA = LA + \pi r^2$
   $SA = $ _____

4. square pyramid with base
   area 4 ft², slant height 3 ft,
   and length of base 1 ft
   $LA = 4(\frac{1}{2}bh)$
   $LA = $ _____
   $SA = LA + \text{base area}$
   $SA = $ _____

5. sphere with diameter of 20 in.
   $SA = 4\pi r^2$
   $SA = $ _____

Solve.

6. Mr. Matthews challenged his math class to make a model of a cylinder out of cardboard. He wants the cylinder to have the following measurements: height = 6 in.; diameter = 3 in. How many square inches of cardboard will each student need to make the model? _____

7. Bill bought a 50-sq-ft roll of wrapping paper. He wants to use the paper to wrap a present in a box with a length of 2 ft, width of 1.5 ft, and height of 0.4 ft. What is the maximum amount of paper Bill will have left, if he wraps the entire box? _____

8. How much material is used to make a ball with a diameter of 6 inches? _____

9. Little Johnny's birthday hat is shaped like a cone. It has a slant height of 4 inches and a diameter of 3 inches. How much cardboard was used to make the hat? _____

10. Little Mary is playing with two building blocks, each 5 inches long, 3 inches wide, and 2 inches high. If she stacks one block directly on top of the other, what is the surface area of the two blocks stacked together? _____

Chapter 9 Volume of Three-Dimensional Shapes

Introduction

Invite students to bring different size containers to class. Discuss the many different shapes and sizes of the containers. Determine which size container is the best size for a particular product. Challenge students to design their own containers for products. Be sure they take into account the size and shape of the container.

Lesson 9•1 Volume of Rectangular Prisms

Using Manipulatives

Use the boxes from Lesson 8•1. Have students work in small groups to find the volume of the boxes. Give students unit cubes to review finding volume by counting cubic units. Tell students to pick one of the smaller boxes, label it Box A, and then fill it with one layer of cubes. Ask students how to find how many cubes there are without counting all the cubes. [Multiply the number of cubes in one layer by the number of layers; that is, by the height.] Next, ask students to fill one row of cubes to the top of the box. Challenge students to find the total number of cubes needed to fill the box, again, without counting the cubes. [Multiply the number of cubes in one layer by the number of layers; that is, by the height.] Finally, request that students check their answers by filling up the box with the cubes, then counting how many cubes were needed. Filling the entire box with cubes needs to be done only once, to check answers; then the volume can be calculated by using the formula $V = lwh$. As an extension, ask students to compare the volume and the surface area of each box, keeping a written record of the box number, volume, and surface area.

Final Check

Draw a rectangular prism and a cube with the following dimensions on the board:

Rectangular prism: length = 9 in.;
width = 12 in.; height = 4 in.

Cube: length of edge = 12.5 in.

Let students estimate the volume of each figure and then calculate the exact volume.

[rectangular prism: estimate: 400 in.^{3}, exact: 432 in.^{3}; cube: estimate: 2,000 in.^{3}, exact: 1,953.125 in.^{3}]

Lesson 9•2 Volume of Cylinders

Chalkboard Activity

Review what students have learned about cylinders up to this point. Invite volunteers to draw a cylinder on the board, label its height and diameter, and make up dimensions for each. Then have the class find the lateral area and surface area of the cylinder on the board.

Using Manipulatives

Use the containers from Lesson 8•2. Have students work in small groups to find the volume of the containers. Ask students to compare the volume and surface area of each container.

Error Analysis

Some students may make careless errors in calculations when completing exercises 10 to 15. Remind students that the radius is used to find the area of a circle. If the diameter is given, it must be divided by 2. Also, stress that volume is measured in cubic units. Encourage students to work slowly and carefully, checking each calculation as they complete the exercises.

Lesson 9•3 Volume of Rectangular Pyramids

Alternate Teaching Approach

You will need a plastic pyramid container and a rectangular prism container with the same base area and height. Ask students to write the formula for the volume of the rectangular prism. Then demonstrate how the volume of the pyramid compares to the volume of the prism by filling the pyramid with water and then pouring it into the prism. Have students keep track of the number of times you need to fill the pyramid and empty it before the prism is full. When they determine that you filled the pyramid three times, ask them to write a formula for the volume of the pyramid. They should conclude that the volume of the pyramid is one third the volume of the prism. You may want to have students complete this investigation in small groups.

Error Analysis

Be sure students use the correct parts of a pyramid when finding its volume. Emphasize that they should use the height of the pyramid to find the volume, not the slant height. You may wish to have students draw several pyramids and then label the height and slant height to emphasize the difference between them.

Lesson 9•4   Volume of Cones

Using a Calculator

Students can use calculators to find the volume of cones. The three methods shown below can be used to find the volume of the cone in the example on page 226.

1 [÷] 3 [X] 3.14 [X] 3 [x^2] [X] 8 [=] [75.36]

1 [÷] 3 [X] [π] [X] 3 [x^2] [X] 8 [=]
[75.398224 ≈ 75.40]

[(] 3.14 [X] 3 [x^2] [X] 8 [)] [÷] 3 [=] [75.36]

Point out the different answer obtained when using the built-in constant π rather than entering 3.14.

Error Analysis

Challenge students to find the volume of a cone with radius 3 feet and height 6 feet. [18π ft^3] Tell students to keep π in their answers, not to substitute 3.14 for π. Check results for the following common student errors.

1. $V = \frac{1}{3}\pi r^2 h = \frac{1}{3}\pi 3^2 \cdot 6 = \frac{1}{3}\pi 6 \cdot 6 = 12\pi$ ft^3

[The error is multiplying the base 3 times the exponent 2 instead of multiplying 3 by itself to get 9.]

2. $V = \frac{1}{3}\pi r^2 h = \frac{1}{3}\pi 3^2 \cdot 6 = \pi 3 \cdot 2 = 6\pi$ ft^3

[The error is dividing the product $3^2 \cdot 6$ by 3 twice instead of once. Students do this because the 3 divides both factors 3^2 and 6 evenly.]

3. $V = \frac{1}{3}\pi r^2 h = \frac{1}{3}\pi 3^2 \cdot 6 = \frac{1}{3}\pi 18^2 = 108\pi$ ft^3

[The error is multiplying the radius 3 times the height h, and squaring the product, instead of squaring the radius first, getting 9, and then multiplying times the height.]

Lesson 9•5   Volume of Spheres

Alternate Teaching Approach

Provide groups of students with a round balloon to act out the example on page 228. Inform students that they will need to find the approximate diameter of the balloon once it is blown up. Suggest that they use a piece of string to find the approximate

circumference of the balloon. Remind them that after they find the diameter, they will need to find the radius before using the formula. Invite groups to share and compare their results.

Final Check

Challenge students to use mental math to determine by what factor the original volume of a 10-ft diameter sphere is multiplied to obtain the new volume when the diameter is doubled.

[$V_{old} = \frac{4}{3}\pi 5^3$, $V_{new} = \frac{4}{3}\pi 10^3$, $\frac{V_{old}}{V_{new}} = \frac{10^3}{5^3} = 2^3 = 8$. The new volume is 8 times the original volume.]

Lesson 9•6   Distinguishing Between Surface Area and Volume

Using Manipulatives

Students can work in groups to make nets of a cube and a prism. They can measure and label the area of each face to show the surface area. For volume, groups can form one of the nets into a cube or a prism, calculate the volume, and label the completed shape. If square-inch or square-centimeter cubes are available, invite students to model volume by filling their cubes or prisms with the cubes, counting the number of cubes used to check whether their calculations are accurate.

Working with ESL/LEP Students

Show students an empty box. Ask them whether the amount of cardboard used to make the box represents the surface area or volume of the box. [surface area] Then fill the box with some type of packing material and ask students whether the packing material represents the surface area or volume of the box. [volume] Ask students to describe other examples of surface area and volume.

Final Check

To assess students' understanding of area and volume, discuss the following questions.

1. What is area? [The measure of the surface covered by a polygon.]

2. What is surface area? [The total area of the outside faces of a three-dimensional figure.]

3. What is volume? [The amount of space enclosed by a three-dimensional object.]

4. How is the surface area of a cube different from the volume of the cube? [The surface area is the total area of the outside faces of the cube; the volume is the space inside the cube.]

Name _____ Date _____

Chapter 9   Test

Label each figure with given dimensions. Then find the volume of each figure described. Show your work. (If necessary, round answers to the nearest hundredth.)

1. rectangular prism with length 12 in., width 8.5 in., height 2.5 in.

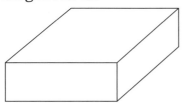

_____

2. cube with edge 11 cm

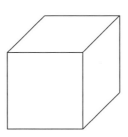

_____

3. cylinder with diameter 8 m and height 12 m

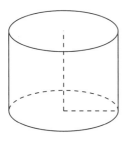

_____

Find the volume of each of the following figures.

4. rectangular pyramid with base area 36 in.² and height 9 in.

_____

5. cone with radius 9.5 cm and height 4.5 cm _____

6. sphere with diameter 4.2 in. _____

7. How many cubic inches of dog food are in a can with a radius of 1.5 in. and a height of 5.5 in.? _____

A cone–shaped paper cup is shown at the right.

8. How many cubic centimeters of water does the cup hold? _____

9. How many cubic centimeters of water will the cup have if it is filled to the dotted line? _____

10. Suppose the cup is filled with water up to the dotted line. If you poured 70 cubic centimeters more water into the cup, will the cup overflow? Explain.

_____

_____

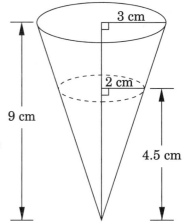

Answers

★ Accept other reasonable answers when a possible answer or drawing is indicated.

Chapter 1 Introduction to Basic Elements

1•1 Points, Lines, Planes

1. Point *M* and point *N*; a line is named by two points on the line.
2. $\overleftrightarrow{PQ}$ 3. $\overleftrightarrow{EF}$ 4. $\overleftrightarrow{ZW}$ 5. $\overleftrightarrow{BC}$
6. Points *A*, *B*, and *C*; plane *ABC*
7. Points *H*, *I*, and *J*; plane *HIJ*
8. Points *Q*, *R*, and *S*; plane *QRS*
9. Points *T*, *U*, and *V*; plane *TUV*
10. Drawing should show a triangle with the three points labeled. **a.** Possible answer: Point *E*, point *F*, point *G*; plane *EFG* **b.** three **c.** Possible answer: $\overleftrightarrow{EF}$, $\overleftrightarrow{FG}$, $\overleftrightarrow{EG}$
11. Drawing should show a four-sided plane with the corners labeled and two dotted diagonal lines inside the plane. **a.** Possible answer: Point *F*, point *G*, point *R*, point *S*; plane *FGRS* **b.** six **c.** Possible answer: $\overleftrightarrow{FG}$, $\overleftrightarrow{GR}$, $\overleftrightarrow{RS}$, $\overleftrightarrow{SF}$, $\overleftrightarrow{FR}$, $\overleftrightarrow{GS}$
12. No, points *A*, *B*, and *D* are coplanar but not collinear; they are in the same plane, but they cannot be connected with a straight line.
13. $\overleftrightarrow{DF}$
14. Possible drawing:

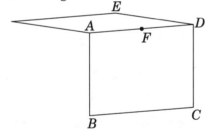

   a. Possible answer: *ABCDF*
   b. Possible answer: Points *B*, *A*, and *F*
   c. Possible answer: Points *A*, *F*, and *D*
15. Drawing should show the cage with points *A*, *B*, and *C* on the front corners and point *D* on one of the rear corners. **a.** Yes, plane *ABC*
   b. Plane *ABC*, plane *BCD*, plane *ACD*
16. Possible answer: Starting at upper left corner:

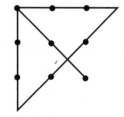

17. Drawing should show two points with a line drawn through them. No, a second line would simply duplicate the first line.
18. Drawing should show two lines intersecting at one point. Two straight lines can only intersect at one point.

1•2 Line Segments and Rays

1. Yes, the endpoint is *N*. The endpoint is always named first.
2. Point *R*, point *S*
3. Drawing should show $\overrightarrow{GF}$ with *G* as the endpoint.
4. Drawing should show $\overline{MN}$; endpoints are *M* and *N*.
5. Ray *FG*
6. Ray *MN*
7. Line segment *XY* (or *YX*)
8. Drawing should show two patterns of line segments connecting Arianne with each of her friends once. There should be numbers by each name to show the order of the connections. Line *segments* must be used, not rays. Each path between friends has two endpoints; rays have only one endpoint.
9. Possible answer: A line, a line segment, and a ray are alike because they are all straight and all are made up of two or more points. They are different because lines have many points and extend in opposite directions without ending, but a line segment has two definite endpoints, and a ray has one endpoint and extends in only one direction.

1•3 Angles

1. Yes; when three letters are used to name an angle, the middle letter (*I*) is the letter at the vertex.
2. Yes; the sides of an angle are rays having a common endpoint (*I*), which is the vertex of the angle.
3. Drawing in text shows an angle with sides $\overrightarrow{GF}$ and $\overrightarrow{GH}$ and vertex *G*.
4. Drawing should show an angle with sides $\overrightarrow{ML}$ and $\overrightarrow{MP}$ and vertex *M*.
5. Drawing should show an angle with sides $\overrightarrow{JH}$ and $\overrightarrow{JK}$ and vertex *J*.
6. Drawing should show an angle with sides $\overrightarrow{XW}$ and $\overrightarrow{XY}$ and vertex *X*.
7. c **8.** e **9.** d
10. a **11.** b

12. Drawing should show examples of angles in the classroom, in such things as doorways, windows, desks and chairs, items on the bulletin board, etc.

13. Possible answer: Students who missed class today should learn the definitions of *angle*, *sides*, and *vertex*. They should also learn how an angle and its parts are named and be able to draw and label an angle if given its name.

1•4   Parallel and Perpendicular Lines

1. Marie could tell from the names of lines *CD* and *CB* that the lines had one point in common, point *C*. This meant that the lines intersected, so they could not be parallel.

2. Drawing in text shows lines *RS* and *TV* intersecting at point *M*.

3. Drawing should show lines *XY* and *ZY* intersecting at point *Y*. Both lines should extend in two directions after intersecting.

4. Drawing should show lines *RS* and *TU* intersecting at point *V*.

5. Drawing should show line *KL* with a point near the middle, *I*, where line *IJ* begins.

6. Line *FG* is parallel to line *JK*. $\overleftrightarrow{FG} \parallel \overleftrightarrow{JK}$

7. Line *PY* is parallel to line *MZ*. $\overleftrightarrow{PY} \parallel \overleftrightarrow{MZ}$

8. Line *WR* is parallel to line *SX*. $\overleftrightarrow{WR} \parallel \overleftrightarrow{SX}$

9. Line *QL* is parallel to line *MN*. $\overleftrightarrow{QL} \parallel \overleftrightarrow{MN}$

10. b

11. a

12. d

13. c

14. Drawing should show a pair of perpendicular lines intersecting. Possible answer: The square edges of a book or a piece of paper

15. Possible answer: $\overleftrightarrow{AB}$ and $\overleftrightarrow{AD}$, $\overleftrightarrow{BD}$ and $\overleftrightarrow{EC}$

16. Possible answer: $\overleftrightarrow{AB} \perp \overleftrightarrow{AD}$ and $\overleftrightarrow{EC} \perp \overleftrightarrow{AD}$

17. $\overleftrightarrow{AB} \parallel \overleftrightarrow{EC}$

18. **a.** Possible answer: Lines to separate lanes of traffic or to show parking places

    b. Possible answer: Lines indicating a crosswalk would be perpendicular to lines separating traffic lanes; also, lines indicating a railroad crossing would be perpendicular to each other.

19. Yes. If one of two parallel lines is perpendicular to a third line, the second parallel line is also perpendicular to the third line. Drawing should show two parallel lines that are perpendicular to a third line.

1•5   Angles Formed by Intersecting Lines

1. Drawing in text shows lines *MN* and *TV* intersecting at point *X*. ∠*MXT* and ∠*VXN*, ∠*MXV* and ∠*TXN*

2. Drawing should show lines *LK* and *GH* intersecting at point *M*. ∠*LMH* and ∠*GMK*, ∠*LMG* and ∠*HMK*

3. Drawing should show lines *CD* and *EF* intersecting at point *P*. ∠*EPC* and ∠*DPF*, ∠*EPD* and ∠*CPF*

4. Drawing should show lines *ST* and *QR* intersecting at point *U*. ∠*SUR* and ∠*QUT*, ∠*SUQ* and ∠*RUT*

5. Possible answer:
    a. ∠3, ∠4, ∠5, and ∠6
    b. ∠1, ∠2, ∠7, and ∠8

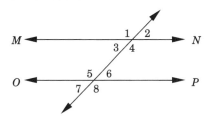

6. ∠1 and ∠6, ∠2 and ∠5, ∠3 and ∠8, ∠4 and ∠7

7. ∠2, ∠6, ∠3, and ∠7

8. ∠1, ∠5, ∠4, and ∠8

9. Possible answer:
    a. ∠2 and ∠8, ∠3 and ∠5
    b. ∠1 and ∠7, ∠4 and ∠6

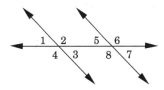

10. ∠2 and ∠7, ∠3 and ∠6

11. ∠1 and ∠8, ∠4 and ∠5

12. Possible drawing:

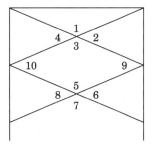

    a. Possible answer: ∠1 and ∠3, ∠2 and ∠4, ∠5 and ∠7, ∠6 and ∠8

    b. Possible answer: ∠3, ∠9, ∠5, and ∠10 **c.** Possible answer: ∠2 and ∠9, ∠9 and ∠6, ∠8 and ∠10, ∠10 and ∠4

    d. Possible answer: ∠1, ∠2, ∠4, ∠6, ∠7, and ∠8

    e. Possible answer: ∠2 and ∠4, ∠6 and ∠8

13. Possible answer: Draw a horizontal line, then two vertical lines that intersect the first line and then intersect each other below the first line.

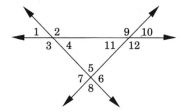

14. An infinite number. If the first two lines are parallel, they extend to infinity without crossing, and they can be crossed by an infinite number of transversals.

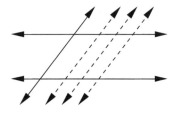

1•6   More Pairs of Angles

1. Adjacent angles are ∠PSQ and ∠QSR, and ∠XWY and ∠YWZ.
2. Linear pair: ∠XWY and ∠YWZ
3. Yes
4. No
5. No
6. Yes
7. Drawing should show 2 separate angles, having no common elements.
8. Drawing should show two adjacent angles having a common side and a common vertex, but their noncommon sides should *not* form a straight line.
9. Drawing should show two adjacent angles whose noncommon sides form a straight line.
10. Possible answer: At 6:00, the minute and hour hands form a straight line; also at 9:16 or 2:44. At 6:00 and 10 seconds, the three hands form two adjacent angles that are a linear pair; also at 9:16 and 5 seconds or 2:44 and 35 seconds. Drawing should show the three hands of the clock forming a linear pair of angles.
11. No. In order for two angles to form a linear pair, they must be adjacent. Drawing should show two separate angles with no common elements.

1•7   Intersecting and Parallel Planes

1. a, b, and d   **2.** c   **3.** Parallel planes
4. Intersecting planes   **5.** Intersecting planes
6. Parallel planes   **7.** e   **8.** d   **9.** a   **10.** c   **11.** b
12. *RT*   **13.** *GH*   **14.** *PQ*   **15.** *XZ*   **16.** 3   **17.** 3

18. a. 8   **b.** 8
19. Possible answer: In a Ferris wheel, the braces are planes that all meet at a single line.

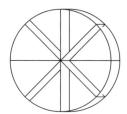

Chapter 1 Review

1. angle   **2.** intersecting   **3.** right   **4.** skew
5. linear pair   **6.** plane   **7.** perpendicular
8. ∠1 and ∠3, ∠2 and ∠4, ∠5 and ∠7, ∠6 and ∠8
9. ∠3 and ∠5, ∠4 and ∠6
10. ∠1 and ∠7, ∠2 and ∠8
11. e   **12.** g   **13.** a   **14.** i   **15.** c   **16.** b   **17.** d
18. h   **19.** f
20. **a.** They are parallel lines.
    b. They are parallel lines.
    c. They are perpendicular lines.
21. Possible answer:
    (1) Two parallel lines intersected by a transversal.

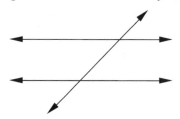

    (2) Three lines that intersect at one point.

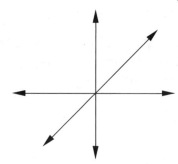

    (3) Three lines intersecting to form a three-sided figure.

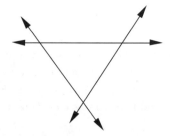

Chapter 1 Practice Test

1. Possible answer: Points *C*, *Y*, and *D*
2. Possible answer: $\overline{AZ}$ and $\overline{ZM}$
3. Possible answer: $\overleftrightarrow{AB}$ and $\overleftrightarrow{NM}$
4. Possible answer: $\overrightarrow{ZB}$ and $\overrightarrow{XR}$
5. Possible answer: ∠*CYZ* and ∠*ZYD*
6. Possible answer: ∠*AZM* and ∠*YZB*
7. Possible answer: ∠*NYC* and ∠*CYZ*
8. Possible answer: ∠*RXC* or ∠*SXC*
9. Possible answer: $\overleftrightarrow{RS}$ and $\overleftrightarrow{XD}$
10. Possible answer: $\overleftrightarrow{AB}$ and $\overleftrightarrow{XD}$
11. Possible answer: $\overleftrightarrow{NM}$ and $\overleftrightarrow{AB}$ (or $\overleftrightarrow{XD}$)
12. Possible answer: $\overleftrightarrow{MN}$
13. Possible answer: ∠*AZY* and ∠*ZYD*, or ∠*CYZ* and ∠*YZB*
14. Possible answer: ∠*AZM* and ∠*NYD*, or ∠*CYN* and ∠*MZB*
15. Possible answer: $\overleftrightarrow{XD}$ and $\overleftrightarrow{RS}$
16. Possible answer: *C, Y, D*
17. Possible answer: The base and lower floors of the skyscraper are slightly larger than the upper floors because the base and lower floors must support the floors above them. This causes the vertical sides to slant slightly toward each other.
18. Parallel
19. Possible answer: Acute angles; if the sides of the building slant slightly inward, these angles would actually be a little less than 90°.
20. Diagram for exercise 17 should show a skyscraper with the sides leaning slightly inward and the floors shown as a series of parallel planes. Pictures for exercise 18 should show the planes and angles of the floors of a skyscraper.

Chapter 1 Test

1. Possible answer: *L, Y,* and *R*
2. Possible answer: $\overline{LY}$ and $\overline{YR}$
3. Possible answer: $\overleftrightarrow{YL}$ and $\overleftrightarrow{YR}$
4. Possible answer: $\overrightarrow{YQ}$ and $\overrightarrow{YR}$
5. Possible answer: ∠*YXT* and ∠*TXP*
6. Possible answer: ∠*YXT* and ∠*SXP*
7. Possible answer: ∠*QYL* and ∠*QYX*
8. Possible answer: ∠*VPW*
9. Possible answer: $\overleftrightarrow{LW}$ and $\overleftrightarrow{VP}$
10. Possible answer: $\overleftrightarrow{QR}$ and $\overleftrightarrow{ST}$
11. Possible answer: $\overleftrightarrow{LW}$
12. Possible answer: ∠*RYX* and ∠*SXY*
13. Possible answer: ∠*LYQ* and ∠*TXP*
14. Possible answer: *y, x,* and *p*
15. Possible answer: $\overleftrightarrow{LW}$ and $\overleftrightarrow{VP}$

16. Possible answer: It would use the least amount of material for construction.
17. Possible answer: It does not have the shape of a rectangle.
18. Possible answer: Depending on how wide the door is open, it may be acute, right, or obtuse.
19. Answers will vary but student drawing should show sets of parallel lines with two parallel transversals.
20. Answers will vary but student drawing should show that for question 16 a rectangle is formed, and for question 17 the sides of the building slant.

Chapter 2   Angles

2•1   Classifying Angles

1. Yes, an acute angle must be greater than 0° and less than 90°.
2. acute 3. obtuse 4. straight 5. right
6. obtuse 7. acute
8. Drawing should show an acute angle.
9. Drawing should show an obtuse angle.
10. Drawing should show an angle of 90°.
11. Drawing should show an angle of 180°.
12. Obtuse. The sum equals 180°, so if one angle is less than 90°, the other must be more than 90°.
13. none 14. ∠*LHJ* 15. none 16. 6
17. ∠*SMR*, ∠*ZMT*
18. ∠*SMT*, ∠*TMR*, ∠*RMZ*, ∠*ZMS*
19. ∠*SMV*, ∠*VMT* 20. ∠*ZMV*, ∠*VMR*
21. Possible answer: ∠*CGF*
22. Possible answer: ∠*CDA*
23. Possible answer: ∠*CAD*
24. Possible answer: ∠*AGB*
25. **a.** right angles
    b. Perpendicular lines intersect at right angles.

2•2   Recognizing Complementary and Supplementary Angles

1. Yes. A right angle has a measure of 90°. 90° + 90° = 180°, so the angles are supplementary.
2. complementary 3. supplementary
4. supplementary 5. complementary
6. c 7. a 8. b
9. Yes. 44 + 46 = 90. Two angles whose measures have a sum of 90° are complementary angles.
10. Drawing should show two adjacent angles whose measures have a sum of 90°.
11. Drawing should show two angles whose measures have a sum of 180°.
12. 55° 13. 65° 14. ∠*A* and ∠*B*

15. ∠A and ∠C, ∠B and ∠D
16. Possible answer: ∠CAE and ∠EAB
17. Possible answer: ∠CAE and ∠AEB
18. a. No. An acute angle must measure less than 90°, so two acute angles would not have the necessary 180° to be supplementary.
  b. Drawing should show two acute angles whose measures have a sum of clearly less than 180°.

2•3 Using a Protractor to Measure Angles

1. m∠ABC = 40° **2.** m∠M = 110°
3. m∠S = 145° **4.** m∠G = 130°
5. m∠T = 40° **6.** m∠K = 90°
7. a. 70° **b.** 80° **c.** 100° **d.** 110°
8. Possible answer: The angles appear to have different measures, because their positions are different and ∠A has longer sides. However, using a protractor shows that both angles have a measure of 46°.

2•4 Using a Protractor to Draw Angles

1. Drawing in text shows m∠ABC = 100°.
2. Drawing should show m∠RST = 80°.
3. Drawing should show m∠LMN = 150°.
4. Drawing should show m∠HIJ = 95°.
5. Drawing should show m∠QRS = 20°.
6. Drawing should show m∠WXY = 170°.
7. a. Possible answer: Put the center point of a protractor on point E and line up the 0° line on the inner scale of the protractor with the top leg of ∠DEF. Mark the point on the inner scale labeled 40. Remove the protractor and draw a line from point E to the 40° mark. This is the uncommon side of your new 40° angle.
  b. Drawings should show m∠DEF = 40° and an adjacent angle of 40°.
8. Possible answer:

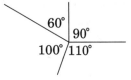

9. a. Possible answer: Place the center point of a protractor on the vertex of the right angle and line up the 0° line on the inner scale of the protractor with the lower leg of the angle. Mark the point on the inner scale labeled 30. Remove the protractor and draw line from the vertex to the 30° mark. This divides the right angle into angles of 30° and 60°.

b. Drawings should show a 90° angle divided into angles of 30° and 60°.

2•5 Recognizing Congruent Line Segments and Angles

1. $\overline{TR} \cong \overline{HJ}$ **2.** ∠GTR ≅ ∠MQW
3. $\overline{LR} \cong \overline{OK}$ **4.** ∠AXC ≅ ∠PTV
5. $\overline{AX} \cong \overline{XB}$, $\overline{DX} \cong \overline{XC}$, $\overline{XB} \cong \overline{XD}$, $\overline{AC} \cong \overline{BD}$
6. ∠AXB ≅ ∠DXC, ∠BXC ≅ ∠DXA
7. Possible answer: $\overline{AB}$ and $\overline{BC}$ at the top of the trellis appear to be congruent. To check this, I would measure both line segments.
8. 90°. The measure of the two angles must total 180°, and if they are congruent, they must each be 90°.

2•6 Bisecting Line Segments

1. Point T is the midpoint of $\overline{JK}$ and $\overline{RS}$. This is true because point T divides both line segments equally.
2. Point H **3.** Point R **4.** Point Y **5.** Point E
6. $\overline{FH} \cong \overline{HG}$ **7.** $\overline{PR} \cong \overline{RQ}$ **8.** $\overline{XY} \cong \overline{YW}$ **9.** $\overline{CE} \cong \overline{EB}$
10. Answer should include the idea that the circles or arcs are the same distance from each endpoint of the segment.
11. A **12.** B **13.** more than $\frac{1}{2}$ the length of $\overline{AB}$
14. Drawing should show arcs intersecting above and below $\overline{AB}$, and a line drawn between the points where the arcs intersect, intersecting $\overline{AB}$ at point P. $\overline{AP} \cong \overline{RT}$
15. Drawing should show arcs intersecting above and below $\overline{RT}$, and a line drawn between the points where the arcs intersect, intersecting $\overline{RT}$ at point P. $\overline{RP} \cong \overline{PT}$
16. Drawing should show arcs intersecting above and below $\overline{LJ}$, and a line drawn between the points where the arcs intersect, intersecting $\overline{LJ}$ at point P. $\overline{LP} \cong \overline{PJ}$
17. Drawing should show arcs intersecting above and below $\overline{XY}$, and a line drawn between the points where arcs intersect, intersecting $\overline{XY}$ at point P. $\overline{XP} \cong \overline{PY}$
18. Possible answer: Tia can draw two arcs of more than half the length of the wire intersecting above and below the wire, with a line drawn through the points where the angles intersect, intersecting the wire at its midpoint. Then she could show how the same method could be used to bisect each half of the wire, resulting in four pieces of equal length.
19. Establish a midpoint on the top and bottom sides of the square. Connecting these midpoints will divide the square into two rectangles of the same size.

2•7 Bisecting Angles

1. **a.** 45°
   b. A right angle has a measure of 90°, and half of 90° is 45°.
2. $\overrightarrow{SV}$ **3.** $\overrightarrow{OP}$ **4.** $\overrightarrow{DF}$ **5.** $\overrightarrow{MP}$
6. $\angle RSV \cong \angle VST$ **7.** $\angle NOP \cong \angle POQ$
8. $\angle GDF \cong \angle FDE$ **9.** $\angle LMP \cong \angle PMN$
10. 15.5°; Drawing should show $\angle ABC$ with a measure of 62° and $\overrightarrow{BD}$ bisecting the angle.
11. Drawing should show a ray bisecting $\angle ABC$ into two 50° angles.
12. Drawing should show a ray bisecting $\angle PQR$ into two 30° angles.
13. Drawing should show an arc intersecting both sides of $\angle DJF$, two arcs intersecting in the middle of the angle, and a midpoint drawn from that intersection to point J, bisecting $\angle DJF$ into two 37° angles.
14. Drawing should show an arc intersecting $\angle BAC$, two arcs intersecting in the middle of the angle, and a midpoint drawn from that intersection to point A, bisecting $\angle BAC$ into two 70° angles.
15. Possible answer: Use a protractor to measure the angle and divide it in half. Drawing should show four line segments all perpendicular to each other and two additional line segments perpendicular to each other and forming 45° angles with the other four line segments.
16. Acute. An acute angle is an angle of less than 90°, so if it is bisected, the two angles formed will also be less than 90°, making them both acute.
17. Possible answer: A protractor. It is usually easier and faster to use.

2•8 Using Parallel Lines

1. $\angle 3$ and $\angle 5$, $\angle 4$ and $\angle 6$
2. $\angle 2$ and $\angle 6$, $\angle 3$ and $\angle 7$, or $\angle 4$ and $\angle 8$, $\angle 1$ and $\angle 5$
3. $\angle 3 \cong \angle 5$, $\angle 2 \cong \angle 8$
4. $\angle 1 \cong \angle 5$, $\angle 2 \cong \angle 6$, $\angle 4 \cong \angle 8$, $\angle 3 \cong \angle 7$
5. **a.** 140° **b.** 40° **c.** 40° **d.** 140° **e.** 140° **f.** 40°
6. 80° **7.** 100° **8.** 80° **9.** 80° **10.** 100° **11.** 100°
12. No. $\angle 5$ and $\angle 6$ are supplementary angles.
13. $\angle 1$ and $\angle 6$, $\angle 2$ and $\angle 5$, $\angle 3$ and $\angle 8$, $\angle 4$ and $\angle 7$
14. Possible answer: $\angle 1$ and $\angle 5$, $\angle 2$ and $\angle 3$
15. 105° **16.** 105° **17.** 125° **18.** 125° **19.** 105°
20. 55° **21.** 75° **22.** 105° **23.** 55° **24.** 75°
25. 125° **26.** 55° **27.** 75° **28.** 125°
29. **a.** Possible answer: AB, CD, and EF
    b. Possible answer: GH and IJ
    c. Possible answer: $\angle 5$ and $\angle 10$, $\angle 6$ and $\angle 9$, $\angle 7$ and $\angle 12$, $\angle 8$ and $\angle 11$, $\angle 13$ and $\angle 18$, $\angle 14$ and $\angle 17$, $\angle 15$ and $\angle 20$, $\angle 16$ and $\angle 19$
    d. Possible answer: $\angle 1$ and $\angle 9$, $\angle 2$ and $\angle 10$, $\angle 3$ and $\angle 11$, $\angle 4$ and $\angle 12$, $\angle 13$ and $\angle 21$, $\angle 14$ and $\angle 22$, $\angle 15$ and $\angle 23$, $\angle 16$ and $\angle 24$
    e. Possible answer: $\angle 1$ and $\angle 6$, $\angle 2$ and $\angle 5$, $\angle 3$ and $\angle 8$, $\angle 4$ and $\angle 7$, $\angle 9$ and $\angle 14$, $\angle 10$ and $\angle 13$, $\angle 11$ and $\angle 16$, $\angle 12$ and $\angle 15$, $\angle 17$ and $\angle 22$, $\angle 18$ and $\angle 21$, $\angle 19$ and $\angle 24$, $\angle 20$ and $\angle 23$ **f.** Possible answer: $\angle 1$ and $\angle 2$, $\angle 5$ and $\angle 6$, $\angle 1$ and $\angle 5$, $\angle 2$ and $\angle 6$
30. Yes, if the transversal is perpendicular to both parallel lines. See drawings.

Chapter 2 Review

1. degrees **2.** acute **3.** bisector **4.** measure
5. complementary **6.** supplementary **7.** straight
8. protractor **9.** bisect **10.** compass **11.** midpoint
12. 35°; acute **13.** 110°; obtuse **14.** 90°; right
15. Drawing should show an angle of 80° bisected into two angles of 40°.
16. Drawing should show $\overline{FG}$ with the midpoint marked.
17. 55° **18.** 55° **19.** 55° **20.** 125° **21.** 125°
22. 125° **23.** 125°
24. Possible answer: Using a ruler or a compass and straightedge, Miriam should measure the board and find the midpoint. This will tell her where to cut the board to get two equal pieces.
25. Possible answer: Timi could use a protractor to measure the size of the angle formed by the piece of pie. Then she could divide the number of degrees by 2 to determine what size half of the angle would be. Using a protractor, she could determine where a bisector would divide the piece of pie into two equal pieces.
26. Possible answer: Z contains 2 congruent angles because it contains 2 parallel lines cut by a transversal, so the alternate interior angles are congruent. Capital F contains 2 congruent angles because the crossbar of the F is perpendicular to the back of the F, resulting in two adjacent right angles. Drawing should show the letters that have been chosen, with the congruent angles marked.

Chapter 2 Practice Test

1. $\angle XOY$ or $\angle WOZ$
2. Possible answer: $\angle ROX$
3. Possible answer: $\angle SOX$
4. Possible answer: $\angle WOR$ and $\angle ROX$
5. Possible answer: $\angle WOR$ and $\angle ROZ$
6. Possible answer: $\angle WOS$ and $\angle WOR$
7. 120° **8.** 60° **9.** 120° **10.** 60° **11.** 60°
12. 120° **13.** 60°

14. Drawing should show a midpoint equidistant from points *D* and *W*.

15. Drawing should show the angle bisected to form two 25° angles.

16. **a.** ∠1 ≅ ∠2, ∠3 ≅ ∠4, ∠5 ≅ ∠6 **b.** Possible answer: If you can predict the angles the ball will take, then you can better plan how to hit the ball.

Chapter 2 Test

1. Possible answer: ∠*ACB*
2. Possible answer: ∠*ECD*
3. Possible answer: ∠*ACD*
4. Possible answer: ∠*ACD* and ∠*DCE*
5. Possible answers: ∠*ACD* and ∠*CEF* or ∠*ECD* and ∠*BEF*
6. 135° **7.** 45° **8.** 45° **9.** 135°
10. 45° **11.** 135° **12.** 45°
13. Possible drawing: Point *X* should be midpoint of $\overline{GK}$.

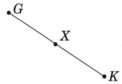

14. Possible drawing: Ray $\overrightarrow{XZ}$ divides ∠*WXY* into two congruent angles.

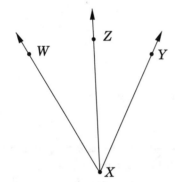

15. **a.** ∠*ABE* and ∠*CBD* **b.** Possible answer: Knowing this allows scientists to accurately predict results or how to adjust equipment to better collect light.

Chapter 3   Triangles

3•1   Identifying the Line Segments that Form Triangles

1. Δ*WXY*; Sides: $\overline{WX}$, $\overline{XY}$, $\overline{YW}$; Vertices: Points *W*, *X*, *Y*
2. Drawing should show Δ*FGH*, with point *P* in the interior of the triangle and point *Q* in the exterior.
3. **a.** $\overline{DF}$, $\overline{FN}$, $\overline{ND}$ **b.** Points *D*, *F*, *N*

4. **a.** Drawing should show Δ*LMN*.
   b. Δ*LMN*

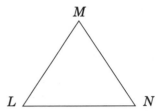

5. Possible answer: tripods for the cameras of photographers, for stability in bridge construction, in roofs of houses for stability and water runoff
6. Possible answer: Yes, though the three segments might form a smaller triangle

3•2   Classifying Triangles

1. Yes. An isosceles triangle has at least two congruent sides. Since Roberto's triangle had three congruent sides, it was also an equilateral triangle.
2. Scalene **3.** Equilateral **4.** Isosceles
5. isosceles; 5 m **6.** isosceles; cannot tell from illustration; 4 in. *4 in equil.*
7. Drawing must be a triangle with at least two congruent sides.
8. Drawing must be a triangle with all sides having different lengths.
9. equilateral, isosceles, and acute
10. obtuse **11.** right **12.** acute **13.** right
14. acute **15.** obtuse
16. 35°; 65°; 80°; acute **17.** 45°; 90°; 45°; right
18. Possible answer: Triangles in rectangles are acute and obtuse; triangles in squares are right.
19. **a.** Yes, if one of the angles of an isosceles triangle is a right angle, then it is an isosceles right triangle. **b.** Drawings should show a right triangle with two congruent sides.

3•3   Identifying Congruent Triangles

1. *ABC* **2.** *KLM* **3.** *TUV* **4.** $\overline{RT}$ **5.** ∠*S* **6.** $\overline{TS}$
7. ∠*T* **8.** ∠*R* **9.** $\overline{RS}$ **10.** Δ*BAD* ≅ Δ*MND*
11. Δ*QPL* ≅ Δ*RST* **12.** Δ*XYZ* ≅ Δ*YXW*
13. 6 cm; 8 cm; 12 cm
14. Possible answer: They used congruent triangles to assure that the design was even and balanced throughout.

3•4   Proving Triangles Congruent

1. a, b
2. $\overline{EF}$ ≅ $\overline{HI}$, $\overline{FG}$ ≅ $\overline{IJ}$, $\overline{GE}$ ≅ $\overline{JH}$
3. $\overline{RT}$ ≅ $\overline{QX}$, $\overline{RW}$ ≅ $\overline{QZ}$, $\overline{TW}$ ≅ $\overline{XZ}$

4. $\overline{LM} \cong \overline{SR}$, $\overline{LN} \cong \overline{RT}$, $\overline{MN} \cong \overline{ST}$

5. $\overline{RS} \cong \overline{NO}$, $\overline{RT} \cong \overline{NP}$, $\overline{ST} \cong \overline{OP}$

6. Yes; $\triangle ABC \cong \triangle SRT$ **7.** Yes; $\triangle RLK \cong \triangle RMB$

8. Yes; $\triangle ABC \cong \triangle DCB$ **9.** No   **10.** Yes; ASA

11. Yes; SAS   **12.** No   **13.** Yes; ASA

14. $\overline{AB} \cong \overline{TU}$, $\overline{BC} \cong \overline{UV}$, $\angle B \cong \angle U$

15. $\angle I \cong \angle L$, $\overline{IG} \cong \overline{LJ}$, $\angle G \cong \angle J$

16. $\overline{PQ} \cong \overline{FD}$, $\overline{PR} \cong \overline{FS}$, $\overline{QR} \cong \overline{DS}$

17. $\overline{JL} \cong \overline{MP}$, $\angle L \cong \angle P$, $\overline{LK} \cong \overline{PN}$

18. Answers will vary but should include the fact that triangles on each side are congruent.

19. Possible answer: After measuring, use the SAS, SSS, and ASA postulates.

3•5   Angles of a Triangle

1. 180°; 30°; 180°; 80°; 180°; 180°; 80°; 100°

2. 180°; 40°; 180°; 100°; 180°; 180°; 100°; 80°

3. 180°; 180°; 110°; 180°; 180°; 110°; 70°

4. 180°; 28°; 103°; 180°; 131°; 180°; 180°; 131°; 49°

5. 90° Possible answer: Since the right angle measures 90°, the two acute angles must have a sum of 180° − 90° = 90°.

6. 65°   **7.** 30°   **8.** 28°   **9.** 58°   **10.** 100°; 80°; 100°

11. 120°; 60°; 120°   **12.** 75°   **13.** 80°   **14.** 155°

15. 105°   **16.** 55° and 35°

17. Answers will vary but should be similar to the diagrams and questions in exercises 10 to 15.

3•6   Right Triangles

1. Yes. Possible answer: $\overline{MP}$ is the third side, and the two legs have already been named.

2. $\overline{SR}$ and $\overline{TR}$; $\overline{ST}$

3. legs: $\overline{XW}$ and $\overline{WV}$; hypotenuse: $\overline{XV}$

4. $\overline{ZY}$ and $\overline{XY}$; $\overline{XZ}$

5. legs: $\overline{PQ}$ and $\overline{PR}$; hypotenuse: $\overline{QR}$

6. t and v; u

7. n and l; m

8. Possible answer: No. The square of the hypotenuse is equal to the sum of the squares of the two legs. However, the hypotenuse itself is not longer than the sum of the lengths of the legs.

9. 8^2, 6^2, 64; 36; 100; 10

10. 7^2; 10^2; 49; 100; 100; 49; 51; $\sqrt{51}$

11. 8.9   **12.** 8.7   **13.** 15.6   **14.** 10.4   **15.** 18.4

16. 13.2

17. a. Home plate to first base is a; first base to second base is b; home plate to second base is c.
   b. 127.3 ft

18. Possible answer: Yes, because the Pythagorean Theorem is true only for right triangles

3•7   Similar Triangles

1. Yes, because corresponding angles of similar triangles are congruent

2. TVU   **3.** GFH   **4.** DBE   **5.** DFE

6. Drawings should show $\triangle DEF \sim \triangle RQP$. $\overline{DF} \leftrightarrow \overline{RP}$, $\overline{FE} \leftrightarrow \overline{PQ}$, $\overline{ED} \leftrightarrow \overline{QR}$

7. Drawings will vary. $\overline{GH} \leftrightarrow \overline{CD}$, $\overline{GI} \leftrightarrow \overline{CB}$, $\overline{HI} \leftrightarrow \overline{DB}$

8. Drawings should show $\triangle CDE \sim \triangle CAB$. $\overline{CD} \leftrightarrow \overline{CA}$, $\overline{DE} \leftrightarrow \overline{AB}$, $\overline{EC} \leftrightarrow \overline{BC}$

9. Drawings should show $\triangle MNP \sim \triangle QNR$. $\overline{MN} \leftrightarrow \overline{QN}$, $\overline{MP} \leftrightarrow \overline{QR}$, $\overline{NP} \leftrightarrow \overline{NR}$

10. Possible answer: Congruent and similar triangles are alike because they both have equal angles. The difference is that, with congruent triangles, the sides are equal, but with similar triangles, the sides are proportional.

11. $\overline{GH}$; $\overline{HI}$; $\overline{IG}$

12. $\frac{DE}{GH} = \frac{EF}{HI}$; $\frac{EF}{HI} = \frac{FD}{IG}$; $\frac{DE}{GH} = \frac{FD}{IG}$

13. $d = 15$; $e = 12$ **14.** $c = 4$; $d = 5$

15. $m = 10$; $n = 18$ **16.** $WX = 10$; $h = 36$

17. a. 2.5 in.   **b.** Triangle should have two sides each 2.5 in. long and one side 1.5 in. long.

18. Possible answer: Yes, the similarity is based on a ratio of 1 to 1.

3•8   Using Inequalities for Sides and Angles of a Triangle

1. Yes. Possible answer: The sum of the lengths of any two sides is greater than the length of the third side.

2. $\overline{CD} + \overline{DE} > \overline{EC}$; $\overline{DE} + \overline{EC} > \overline{CD}$; $\overline{CD} + \overline{EC} > \overline{DE}$

3. $\overline{MN} + \overline{NO} > \overline{MO}$; $\overline{MO} + \overline{ON} > \overline{MN}$; $\overline{MO} + \overline{MN} > \overline{ON}$

4. $\overline{QS} + \overline{RQ} > \overline{SR}$; $\overline{QS} + \overline{SR} > \overline{RQ}$; $\overline{RQ} + \overline{SR} > \overline{QS}$

5. $\overline{LM} + \overline{LK} > \overline{MK}$; $\overline{MK} + \overline{LK} > \overline{LM}$; $\overline{ML} + \overline{MK} > \overline{KL}$

6. Yes   **7.** No   **8.** Yes   **9.** Yes   **10.** Yes   **11.** No

12. Yes   **13.** Yes   **14.** No

15. Yes. Since the sides are all equal in length, the angles should all have the same measure.

16. $\angle F$; $\angle E$; $\angle D$

17. $\angle Q$; $\angle S$; $\angle R$

18. $\overline{DC}$; $\overline{BC}$; $\overline{DB}$

19. $\overline{GH}$; $\overline{FG}$; $\overline{FH}$

20. $\angle S > \angle R$

21. $\angle G$

22. $\overline{KL}$

23. Possible answer: No. It is at most 8 blocks from school to the library, and at most 10 blocks from the library to the skating path. So Morte could only have walked at most 18 blocks.

24. The angles are congruent because the sides are congruent and the angles are related in the same way as the sides across from them.

Chapter 3 Review

1. equilateral   **2.** size, shape   **3.** scalene
4. corresponding parts   **5.** obtuse   **6.** legs
7. similar   **8.** isosceles
9. $\overline{AB} \cong \overline{XY}$, $\angle A \cong \angle X$, $\overline{AC} \cong \overline{XZ}$
10. $\angle M \cong \angle R$, $\overline{MP} \cong \overline{RT}$, $\angle P \cong \angle T$
11. Yes   **12.** No   **13.** No   **14.** 85°   **15.** 45°   **16.** 140°
17. 40°   **18.** 18.0 m   **19.** 7.9 in.
20. $x = 8$ m; $NM = 3$ m
21. 8 ft   **22.** $\angle J > \angle N$ **23.** $\angle M < \angle J$ **24.** 12 ft
25. Answers should show an understanding of ratios and scale drawings, by using reasonable ratios to make the drawings. A reasonable ratio might be 1 in.= 2 ft.

Chapter 3 Practice Test

1. acute and scalene   **2.** acute and isosceles
3. acute and equilateral   **4.** acute and scalene
5. right and isosceles   **6.** obtuse and scalene
7. $\overline{AC} \cong \overline{AD}$, $\overline{AB} \cong \overline{AB}$, $\overline{BC} \cong \overline{BD}$
8. 28.3 m   **9.** 40°   **10.** 120°   **11.** 140°   **12.** 100°
13. 10   **14.** 20
15. No. For any triangle, the sum of the lengths of any two sides must be greater than the length of the remaining side. In this case, sides of 4 and 8 added together are not greater than 12.

Chapter 3 Test

1. right, scalene   **2.** acute, isosceles
3. acute, equilateral   **4.** obtuse, scalene
5. obtuse, scalene   **6.** acute, scalene
7. $\overline{CD} \cong \overline{CB}$; $\overline{CE} \cong \overline{CA}$; $\overline{DE} \cong \overline{AB}$
8. 13.4 m
9. 50°   **10.** 130°   **11.** 130°   **12.** 100°   **13.** 12   **14.** 6
15. No, the sum of the lengths of two of the sides, for example 1 and 2, is not greater than the length of the third side, 3.

Chapter 4   Polygons

4•1   Types of Polygons

1. sides: $\overline{LM}$, $\overline{MN}$, $\overline{NO}$, $\overline{OP}$, $\overline{PL}$; vertices: *L, M, N, O, P*; diagonals: $\overline{MP}$, $\overline{NP}$
2. sides: $\overline{ST}$, $\overline{TU}$, $\overline{UV}$, $\overline{VW}$, $\overline{WX}$, $\overline{XY}$, $\overline{YZ}$, $\overline{ZS}$; vertices: *S, T, U, V, W, X, Y, Z*; diagonals: $\overline{SU}$, $\overline{SV}$, $\overline{SW}$, $\overline{SX}$,

$\overline{SY}$, $\overline{TV}$, $\overline{TW}$, $\overline{TX}$, $\overline{TY}$, $\overline{TZ}$, $\overline{UW}$, $\overline{UX}$, $\overline{UY}$, $\overline{UZ}$, $\overline{VX}$, $\overline{VY}$, $\overline{VZ}$, $\overline{WY}$, $\overline{WZ}$, $\overline{XZ}$; 20 diagonals for an 8-sided figure

3.
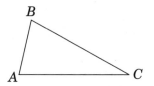

4.
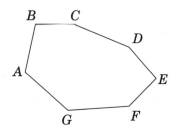

5.
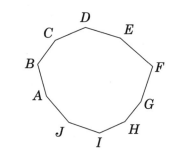

6.
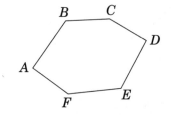

7. Possible answer: Perhaps a blackboard, or floor is tiled—quadrilateral. Watch face may be octagon.
8. The polygon is a non-regular hexagon.

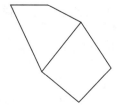

4•2   Properties of Parallelograms

1. a. Rico is correct. Both pairs of opposite sides must be parallel to have a parallelogram. If only one pair of opposite sides are parallel, it may be a trapezoid, not a parallelogram.

   b.

2. sides: $\overline{RS} \cong \overline{TU}$; $\overline{RU} \cong \overline{ST}$; angles: $\angle R \cong \angle T$; $\angle U \cong \angle S$

3. sides: $\overline{LM} \cong \overline{ON}$; $\overline{NM} \cong \overline{OL}$; angles: $\angle L \cong \angle N$; $\angle M \cong \angle O$

4. $\overline{ON} = 8$ cm

5. $\overline{NM} = 5$ cm

6. $m\angle L = 45°$

7. $m\angle M = 135°$

8. a. 7

   b. 3

9. line segments: $\overline{XY} \cong \overline{ZW}$, $\overline{YZ} \cong \overline{WX}$, $\overline{XA} \cong \overline{AZ}$, $\overline{YA} \cong \overline{AW}$; triangles: $\triangle YXW \cong \triangle WZY$, $\triangle XYZ \cong \triangle ZWX$

10. line segments: $\overline{RQ} \cong \overline{TS}$, $\overline{RS} \cong \overline{TQ}$, $\overline{RP} \cong \overline{TP}$, $\overline{QP} \cong \overline{SP}$; triangles: $\triangle RTQ \cong \triangle TRS$, $\triangle RSQ \cong \triangle TQS$

11. line segments: $\overline{JK} \cong \overline{LM}$, $\overline{KL} \cong \overline{MJ}$, $\overline{JN} \cong \overline{LN}$, $\overline{KN} \cong \overline{MN}$; triangles: $\triangle JKL \cong \triangle LMJ$, $\triangle JKM \cong \triangle MLK$

12. She needs 6 sticks, 3 that are 1.5 ft long, 2 that are 1 ft long, and 1 that is 2 ft long. The crossbars are not cut in half, they are glued together at the center of the diagonals. Not all students will realize that the crossbars on a kite will consist of two pieces not four as the diagram indicates.

13. Possible answer: No, if you know the measure of one side of a parallelogram, then you know the measure of the opposite congruent parallel side. But the other pair of congruent, parallel sides could be many unknown values.

14. There are 3 different parallelograms formed by 2 congruent scalene triangles. Place congruent edges next to each other to get 3 different parallelograms.

4•3   Rectangles

1. No, it is not a rectangle; it is not a parallelogram.

2. No, it is not a rectangle; it has one pair of parallel sides, not two.

3. Yes, it is a rectangle. It is a parallelogram with 4 right angles.

4. No, it is not a rectangle. It is not a parallelogram.

5. Yes, the diagonals of a rectangle are congruent and the diagonals of a parallelogram bisect each other. Halves of equals are equal.

6. $\overline{FE} = 6$ in.

7. $\overline{DE} = 8$ in.

8. $\overline{DF} = 10$ in.; $\overline{CG} = \overline{GE} = \overline{DG} = \overline{GF} = 5$ in.

9. 8 ft

10. Yes, the angles opposite two right angles will be congruent and thus 90°. If all angles of a parallelogram are 90°, the parallelogram is a rectangle.

4•4   Rhombuses and Squares

1.

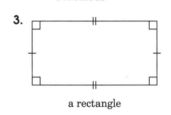

a square

2.

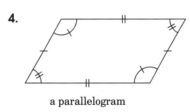

a rhombus

3.

a rectangle

4.

a parallelogram

5. $m\angle 9 = m\angle 10 = m\angle 11 = m\angle 12$, 90°

6. $m\angle 1 = m\angle 2$, 45°   7. $m\angle 3 = m\angle 4$, 45°

8. $m\angle 9 = m\angle 10 = m\angle 11 = m\angle 12$, 90°

9. a. $m\angle 1 = m\angle 2$, 30°

   b. $m\angle KLH = m\angle HJK$, 60°

   c. $m\angle 5 = m\angle 6$, 30°

10. Rhombus $ABCD$; $\triangle CDA \cong \triangle CBA$, $\triangle DCB \cong \triangle DAB$. It appears that $m\angle A = m\angle B = m\angle C = m\angle D = 90°$

11. A parallelogram that has two consecutive sides congruent is a rhombus. Opposite sides of the parallelogram are congruent, so all sides would be congruent.

4•5   Trapezoids

1. A trapezoid is a quadrilateral with exactly one pair of opposite sides parallel, and no sides that are necessarily congruent.

2. Trapezoid; It has two parallel sides but its legs are not congruent.

3. Isosceles trapezoid; It has two parallel sides and its legs are congruent.

4. Trapezoid; It has two parallel sides but its legs are not congruent.

5. Neither: There are no pair of parallel sides.

6. Yes

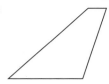

7. No. The sum of the interior angles of a quadrilateral is 360°, so you could not have 4 acute angles in a trapezoid.

8. Yes

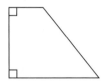

9. No, if 3 sides are congruent, the 4th side would also be congruent.

10. No, consecutive interior angles between the parallel lines are supplementary, which means in a trapezoid only 2 angles can be obtuse.

11. No, congruent parallel bases will automatically make the legs congruent and parallel, forming a parallelogram, not a trapezoid.

12. Not possible. For bases 2 in. and 3 in., the median would measure $\frac{(2+3)}{2} = \frac{5}{2} = 2.5$ in., not 3 in.

13. 15 cm   **14.** 20 cm   **15.** 10.5 in.   **16.** 5 in.

17. 6 cm   **18.** 2 in.

19. Possible answer: Trapezoid $ABCD$; bases: $\overline{AB}$ and $\overline{CD}$, legs: $\overline{AD}$ and $\overline{BC}$, base angles: $\angle D$ and $\angle C$ are a pair and $\angle A$ and $\angle B$ are another pair.

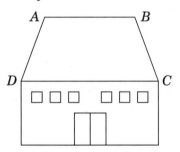

20. Possible answer: $AB = \frac{(17+15)}{2} = \frac{32}{2} = 16$ mm

21. Answers should be about 16 mm ± 2 mm.

22. Yes, the median is parallel to both bases so both trapezoids have one pair of parallel bases. The median divides congruent legs in half, making 2 pairs of congruent legs.

4•6   Regular Polygons

1. No, a rhombus has all 4 sides congruent but is not a regular polygon. All sides are congruent and all angles are congruent in a regular polygon.

2. The second and third figures should be circled.

3. 180°   **4.** 360°   **5.** 540°   **6.** 1,080°

7. 60°   **8.** 90°   **9.** 108°   **10.** 135°

11. 7 sides   **12.** 9 sides

13. Richard is correct. You need to know the sum of the measures of the interior angles. As the following exercises show, the sum of the exterior angles is 360 degrees for all polygons.

14. 120°; 360°   **15.** 90°; 360°   **16.** 72°; 360°

17. 45°; 360°

18. The sum of the exterior angles of a regular polygon is 360°.

19. Sample problem:

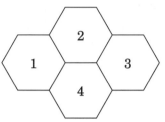

Hexagon 1, 2 and 4 or 2, 3 and 4 have a common vertex. The sum of the angles at that vertex is 360°. Only hexagons which are regular with sides congruent from one hexagon to another could fit together the same way. Non-regular hexagons will not fit together nicely as in the diagram.

4•7   Congruent Polygons

1. $FG \cong PQ$ **2.** $GH \cong QR$ **3.** $HJ \cong RS$

4. $ST \cong JK$ **5.** $TP \cong KF$ **6.** $\angle F \cong \angle P$

7. $\angle G \cong \angle Q$ **8.** $\angle R \cong \angle H$ **9.** $\angle S \cong \angle J$

10. $\angle K \cong \angle T$

11.

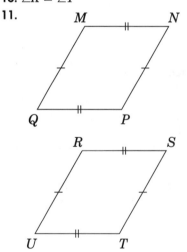

12.

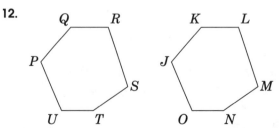

13. Possible answer:

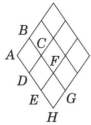

The rhombuses are congruent.
$ABCD \cong EFGH$

14. a. $\angle A \cong \angle E$, $\angle B \cong \angle F$, $\angle C \cong \angle G$, $\angle D \cong \angle H$, $\overline{AB} \cong \overline{EF}$, $\overline{BC} \cong \overline{FG}$, $\overline{CD} \cong \overline{GH}$, $\overline{DA} \cong \overline{HE}$

b. No, two polygons are congruent if they have the same size and shape. However, the polygons do not need to be regular to be congruent.

15. Ned used congruent polygons so the polygons would fit together with no gaps.

4•8   Similar Polygons

1. No, they cannot both be correct. One angle, $\angle D$, must correspond to only one angle, either $\angle J$ or $\angle K$. $\angle F \cong \angle L$ and $\angle G \cong \angle M$. One of the two people is correct, not both.

2. N; O; P; $\overline{NO}$; $\overline{OP}$; $\overline{PL}$, Quadrilateral $RSTV \sim$ Quadrilateral $WXYZ$

3. $\angle S \cong \angle X$, $\angle T \cong \angle Y$, $\angle U \cong \angle Z$, $\angle R \cong \angle W$, $\overline{RS} \leftrightarrow \overline{WX}$, $\overline{ST} \leftrightarrow \overline{XY}$, $\overline{TU} \leftrightarrow \overline{YZ}$, $\overline{UR} \leftrightarrow \overline{ZW}$

4. Hexagon $DEFGHI \sim$ Hexagon $LMNOPQ$, $\angle D \cong \angle L$, $\angle E \cong \angle M$, $\angle F \cong \angle N$, $\angle G \cong \angle O$, $\angle H \cong \angle P$, $\angle I \cong \angle Q$. Both hexagons are regular hexagons so all angles = 120°; $\overline{DE} \leftrightarrow \overline{LM}$, $\overline{EF} \leftrightarrow \overline{MN}$, $\overline{FG} \leftrightarrow \overline{NO}$, $\overline{GH} \leftrightarrow \overline{OP}$, $\overline{HI} \leftrightarrow \overline{PQ}$, $\overline{ID} \leftrightarrow \overline{QL}$

5. No, the polygons must have the corresponding angles congruent also.

6. The first and third sets of figures should be circled.

7. $XY = 12$, $ZY = 10$, $RU = 12$

8. $XW = 16\frac{2}{3}$; $XY = 13\frac{1}{3}$; $2Y = 13\frac{1}{3}$; $AE = 7\frac{1}{5}$

9. $CD = 14$, $KL = 3$, $MN = 3$

10.

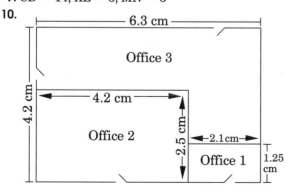

11. Yes, two polygons similar to a third polygon will be similar to each other. Corresponding angles in all three polygons will be congruent. Corresponding sides will be proportional. Drawings will vary.

4•9   Drawing Polygons on the Coordinate Plane

1. Yes, a point on the x-axis has y-coordinate 0.

2. A at $(4, 1)$, B at $(-2, 2)$, C at $(-3, 1)$

3. W at $(4, 1)$, X at $(2, 0)$, Y at $(2, 1)$, Z at $(3, 4)$

4. R at $(-3, 3)$, S at $(0, 1)$, T at $(2, 2)$, U at $(1, 4)$

5. JLK is a triangle.

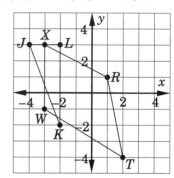

Drawing should include both polygons (5) and (6) on the same coordinate plane given.

7. translation to the right 3 units

8. rotation 180°

9. reflection over x-axis

10. F' at $(6, 3)$, G' at $(7, 1)$, H' at $(3, 2)$

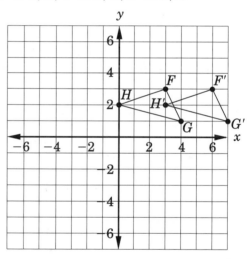

11. F'' at $(3, -3)$, G'' at $(4, -1)$, H'' at $(0, -2)$

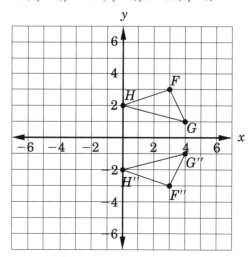

12. Luke should go south to 2nd Ave and West to Oak Lane. Lila's house is in section 28 between *A* and *B* on the map.

13. Reflection image; your left looks like your right side in the mirror.

4•10   Tessellations of Regular Polygons

1. The first and third figures should be circled.

2. equilateral triangles, squares

3. equilateral triangles, squares, rhombuses

4. congruent parallelograms

5. isosceles triangles

6. **a.** Answer 3 is answer 2 plus rhombuses.
   b. The patterns of polygons are repeated.
   c. Tessellations are repeated polygons with no gaps or overlaps.

7. Yes; a tessellation is formed by translating, reflecting, or rotating one or more of the same polygons, which means the sides of one are congruent to many others.

8. Possible answer:

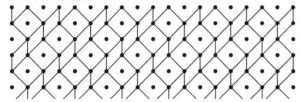

9. Possible answer:

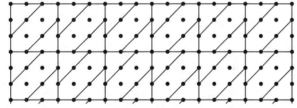

10. Possible answer:

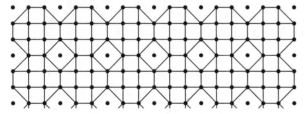

11. Possible answer: The pattern is of repeated polygons. The pattern was formed by tessellating the polygons around one another. Inside, rhombuses were used. See drawing.

12. Possible answer: Clothing and wallpaper are two examples of objects that may display tessellations.

13. No, it is impossible to make a tessellation with a circle because you could not cover an area without gaps or overlapping. Circles are tangent to any polygon or another circle only at one point.

Chapter 4 Review

1. polygon   2. tessellation   3. parallelogram
4. quadrilateral   5. rhombus   6. congruent
7. ordered pair   8. pentagon   9. hexagon
10. triangle   11. quadrilateral   12. rhombus
13. trapezoid   14. square   15. parallelogram
16. 105°   17. 75°   18. 3   19. $\triangle ABD \cong \triangle CDB$
20. 6   21. 4   22. 8
23. $m \angle XQY = m \angle YQZ = m \angle ZQW = m \angle WQX = 90°$
24. **a.** 8
    b. 135°

25.

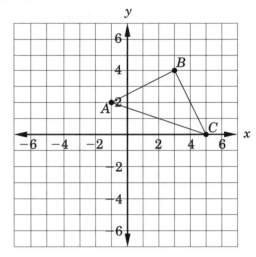

26.

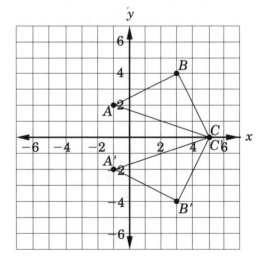

27. Possible answer: Given sides: 3, 4.5, 6, 2.5. Side proportional to 3 will be 5.

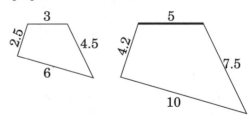

28. Possible answer: Stop sign—octagon. School crossing—pentagon

Chapter 4 Practice Test

1. a 2. b 3. c 4. e 5. f 6. d
7. 100°, 80°, 100° 8. 10 9. 6 sides
10. hexagon 11. 120° 12. 60° 13. rhombus
14. 4 15. 5 16. 10
17.

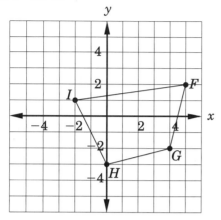

Chapter 4 Test

1. b 2. e 3. f 4. a 5. d 6. c 7. 65°, 115°, 115°
8. 35 9. 5 10. pentagon 11. 108° 12. 72°
13. trapezoid 14. 12 15. 24

Chapter 5 Area and Perimeter of Polygons

5•1 Perimeter of Parallelograms and Rectangles

1. $2(6) + 2(9) = 12 + 18 = 30$ in.
2. $2(12) + 2(8) = 24 + 16 = 40$ m
3. $7 + 7 + 4 + 4 = 22$ cm
4. $2(14) + 2(12) = 28 + 24 = 52$ ft
5. $2(2.5) + 2(5) = 5 + 10 = 15$ m
6. $10.5 + 10.5 + 16 + 16 = 53$ in.
7. 50 ft; 194 ft
8. Possible answers: length = 12 ft, width = 10 ft or length = 14 ft, width = 8 ft. Each combination of lengths and sides must fit the formula $P = 2l + 2w$.

5•2 Area of Parallelograms and Rectangles

1. $A = 6 \cdot 7 = 42$ cm^2
2. $A = 12 \cdot 5 = 60$ ft^2
3. $A = 9 \cdot 4 = 36$ in.^{2}
4. $A = 8 \cdot 11 = 88$ m^2
5. $A = 6.5 \cdot 2.5 = 16.25$ cm^2
6. $A = 9.2 \cdot 13.5 = 124.2$ yd^2
7. 4,700 ft^2

8. Yes. For example, when $l = 6$ and $w = 3$, the perimeter is 18 and the area is 18.

5•3 Estimating Perimeter and Area

1. $P \approx 20$ in.; $A \approx 18$ in.^{2}
2. $P \approx 30$ in.; $A \approx 54$ in.^{2}
3. $P \approx 26$ in.; $A \approx 34$ in.^{2}
4. $P \approx 34$ in.; $A \approx 86$ in.^{2}
5. $P \approx 16$ in.; $A \approx 18$ in.^{2}
6. $P \approx 22$ in.; $A \approx 30$ in.^{2}
7. See drawing. Make sure that the poster meets the size requirements.
8. Place a grid on top of the figure and count the number of complete squares in the figure. Estimate the number of complete squares in the rest of the figure. Then add to get the area.

5•4 Perimeter and Area of a Rhombus or a Square

1. $b = 7$ cm, $h = 6$ cm
2. $b = 9$ in., $h = 8.1$ in.
3. $b = 8.6$ m, $h = 7.5$ m
4. $P = 4 \cdot 6 = 24$ in.; $A = 6 \cdot 5 = 30$ in.^{2}
5. $P = 4 \cdot 12 = 48$ cm; $A = 12 \cdot 10 = 120$ cm^2
6. $P = 4 \cdot 15 = 60$ m; $A = 15 \cdot 13.5 = 202.5$ m^2
7. $P = 6.5 \cdot 7.2 = 26$ m; $A = 7.2 \cdot 6.5 = 46.8$ m^2
8. $P = 4 \cdot 2.5 = 10$ m; $A = 2.5 \cdot 2.25 = 5.625$ m^2
9. $P = 4 \cdot 135 = 540$ yd; $A = 135 \cdot 120 = 16,200$ yd^2
10. Margie is correct. There is no whole number whose square is 32.4, but $6 \cdot 6 = 36$.
11. $P = 4 \cdot 7 = 28$ in.; $A = 7^2 = 49$ in.^{2}
12. $P = 4 \cdot 10 = 40$ cm; $A = 10^2 = 100$ cm^2
13. $P = 4 \cdot 12 = 48$ m; $A = 12^2 = 144$ m^2
14. $P = 4 \cdot 8 = 32$ ft; $A = 8^2 = 64$ ft^2
15. $P = 4 \cdot 4 = 16$ m; $A = 4^2 = 16$ m^2
16. $P = 4 \cdot 1.5 = 6$ yd; $A = 1.5^2 = 2.25$ yd^2
17. 196 ft^2
18. 16 in. 19. 2; Sketches will vary.
20. The rhombus must be a square, because each of its sides has a length of $12 \div 4$, or 3 cm, and $3^2 = 9$, which is its area.

5•5 Properties of the Diagonals of a Rhombus

1. $A = (\frac{1}{2}) \cdot 6 \cdot 10 = 30$ in.^{2}
2. $A = (\frac{1}{2}) \cdot 9 \cdot 15 = 67.5$ cm^2
3. $A = (\frac{1}{2}) \cdot 9 \cdot 6 = 27$ in.^{2}
4. $A = (\frac{1}{2}) \cdot 12 \cdot 16 = 96$ ft^2
5. $A = (\frac{1}{2}) \cdot 7.4 \cdot 4.3 = 15.91$ m^2
6. $A = (\frac{1}{2}) \cdot 84 \cdot 56 = 2,352$ in.^{2}

7. 18 ft

8. Possible answer: 10 m and 20 m; 40 m and 5 m; any two measurements whose product is 200 m².

9. 20.255 in.²

10. a. They are four congruent right triangles.
   b. 4 in.²

11. Yes, a square is a rhombus with four right angles, so the formula can be used.

5•6 Area of Triangles

1. Paula is correct, because the area is half the base, or 8 m, times the height, 5 m, or 40 m².

2. $A = (\frac{1}{2}) \cdot 5 \cdot 4 = 10$ m²

3. $A = (\frac{1}{2}) \cdot 6 \cdot 9 = 27$ in.²

4. $A = (\frac{1}{2}) \cdot 10 \cdot 6 = 30$ cm²

5. $A = (\frac{1}{2}) \cdot 16 \cdot 11 = 88$ ft²

6. $A = (\frac{1}{2}) \cdot 7 \cdot 5 = 17.5$ yd²

7. $A = (\frac{1}{2}) \cdot 8.5 \cdot 2 = 8.5$ m²

8. 5 m

9. Possible answer: 6 km and 8 km; 4 km and 12 km

10. The diagonal of a parallelogram separates it into two congruent triangles. So the area of each triangle is half of the area of the parallelogram. Check drawings.

5•7 Area of Trapezoids

1. Frank is correct; the formula says find half of the height and multiply that times the sum of the lengths of the bases. $(\frac{1}{2})(4)(6 + 8) = 28$.

2. $A = (\frac{1}{2})(2)(2 + 4) = 6$ cm²

3. $A = (\frac{1}{2})(4)(3 + 5) = 16$ m²

4. $A = (\frac{1}{2})(10)(18 + 7) = 125$ in.²

5. $A = (\frac{1}{2})(8)(8 + 11) = 76$ ft²

6. $A = (\frac{1}{2})(9)(6 + 12) = 81$ m²

7. $A = (\frac{1}{2})(4.5)(14 + 8) = 49.5$ yd²

8. 2 congruent right triangles and a rectangle

9. There is no limit to the number of heights you can draw. They all have the same length because they all represent the distance between the two parallel bases.

5•8 Perimeter and Area of Irregular Figures

1. triangle, rectangle   **2.** square, trapezoid

3. 2 trapezoids, 1 rectangle   **4.** rhombus, square

5. $A = (16 \cdot 10) + (\frac{1}{2}) \cdot 16 \cdot 9) = 232$ cm²

6. $A = (144) + [(\frac{1}{2})(14)(4 + 12)] = 256$ in.²

7. $A = [2 \cdot (\frac{1}{2})(2)(3 + 5)] + (5 \cdot 9) = 61$ cm²

8. $A = ((\frac{1}{2}) \cdot 10 \cdot 7) + (12 \cdot 7) = 119$ ft²

9. No, he also needs to know the measures of the two sides of the rectangle.

10. $P = 12 + 12 + 10 + 16 + 10 = 60$ cm

11. $P = 14.5 + 4 + 14.5 + 12 + 12 + 12 = 69$ in.

12. $P = 2 + 3 + 2 + 9 + 2 + 3 + 2 + 9 = 32$ cm

13. $P = 12 + 7 + 12 + 7 + 7 + 7 = 52$ ft

14. about 240 tiles

15. a. 48 ft²   **b.** 26 ft

16. Check drawings. Possible answer:

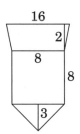

$8(8) + \frac{8 \cdot 3}{2} + [(\frac{1}{2})(16 + 8) \cdot 2] =$
$64 + 12 + 24 = 100$ cm²

17. You can find the perimeter by adding the lengths of all the sides. If you can separate the polygon into polygons whose areas you can find, then you can find the total area.

5•9 Determining the Difference Between Perimeter and Area

1. perimeter   **2.** area   **3.** perimeter   **4.** area

5. area   **6.** perimeter   **7.** area   **8.** perimeter

9. a. 24 ft²   **b.** 24 ft

10. a. 28 ft   **b.** 49 ft²

11. 48 m² describes an area, while 48 m describes a perimeter. Possible answers: For 48 m², a rectangle with length 8 m and width 6 m; for 48 m, a square with sides 12 m

---

Chapter 5 Review

1. area   **2.** length; width   **3.** perimeter

4. $P = 2(9) + 2(6) = 30$ m; $A = 9 \cdot 6 = 54$ m²

5. $P = 2(7) + 2(5) = 24$ in.; $A = 4 \cdot 7 = 28$ in.²

6. $P = 2(13) + 2(10) = 46$ cm; $A = 12 \cdot 10 = 120$ cm²

7. $P = 2(6) + 2(2.5) = 17$ ft; $A = 6 \cdot 2.5 = 15$ ft²

8. $P = 4 \cdot 5 = 20$ ft; $A = (\frac{1}{2}) \cdot 6 \cdot 8 = 24$ ft²

9. $P = 4 \cdot 12.5 = 50$ cm; $A = 12.5^2 = 156.25$ cm²

10. $A = (\frac{1}{2}) \cdot 15 \cdot 6 = 120$ in.²

11. $(2)(3.5) + [(\frac{1}{2})(2)(2.5)] = 9.5$ yd²

12. $4^2 + (8)(5) = 56$ in.²

13. a. 350 ft²   **b.** 106 ft

14. a. 12 in.²   **b.** 72 in.²

15. Answers will vary but should include evidence of research, a diagram of the field and seats, appropriate labels, and an estimate of the area.

Chapter 5 Practice Test

1. 28 cm; 49 cm²
2. 50 in.; 120 in.²
3. 46 m; 126 m²
4. 30 ft; 54 ft²
5. 92.25 cm²
6. 63.25 ft²
7. 2 qt **8.** 54 ft **9.** 5 ft²

Chapter 5 Test

1. 54 in.; 50 in.² **2.** 46 ft; 98 ft² **3.** 84 cm; 441 cm²
4. 24 m; 36 m² **5.** 106.5 in.² **6.** 25.08 cm²
7. 432 ft²
8. No. Two double rolls will cover 400 square feet, but she needs 432 square feet.
9. 54 ft **10.** 106.67 or 107 pieces

Chapter 6 Circles

6•1 Parts of a Circle

1. No, *TW* would be the radius since one endpoint is at the center of the circle and the other is on the circle. Drawing should show ⊙*T* with radius $\overline{TW}$.
2. Drawing should show ⊙*O* with radius $\overline{OW}$.
3. Drawing should show ⊙*R* with radius $\overline{RT}$.
4. Drawing should show ⊙*T* with diameter $\overline{RS}$.
5. Drawing should show ⊙*M* with diameter $\overline{LK}$ and radius $\overline{MQ}$.
6. Drawing should show ⊙*H* with radius $\overline{HK}$ and diameter *BK*.
7. Drawing should show ⊙*J* with diameter $\overline{AB}$ and radius $\overline{JB}$.
8. Kerry is right. Since two times the radius is the diameter, the diameter of Kerry's circle would equal 12 inches. Thus the circles are the same.
9. 6 cm. Drawing should show ⊙*T* with a 6 cm diameter $\overline{SV}$ and a 3 cm radius $\overline{TR}$.
10. 2 cm. Drawing should show ⊙*M* with a 1-cm radius $\overline{MJ}$ and a 2-cm diameter $\overline{LK}$.
11. 2 cm. Drawing should show ⊙P with a 4-cm diameter $\overline{MN}$ and a 2-cm radius $\overline{PV}$.
12. 2.5 cm. Drawing should show ⊙*A* with a 5-cm diameter $\overline{BC}$ and a 2.5-cm radius $\overline{AH}$.
13. Drawing should show a clock with a labeled center and hands. The hands suggest the radii of the circle.
14. A circle has an infinite number of radii and diameters since a radius or diameter can be drawn at any partial degree of the 360° of a circle.

6•2 More Parts of a Circle

1. Yes. Drawing should show $\overline{KF}$ drawn on ⊙*M*.
2. Possible answer: Draw radius $\overline{MP}$. One other arc is *FP*.
3. Possible answer: Draw radius $\overline{MD}$. One other central angle is ∠*LMD*.
4. Possible answer: Draw tangent $\overleftrightarrow{CG}$. One other tangent is $\overleftrightarrow{CG}$.
5. Possible answer: Draw line connecting *L* and *G*. One other secant is $\overleftrightarrow{LG}$.
6. Drawing should show ⊙*O* with chord $\overline{AB}$ and tangent $\overrightarrow{PB}$.
7. Drawing should show ⊙*P* with central ∠*MPN* and secant $\overleftrightarrow{RN}$.
8. Drawing should show ⊙*B* with tangent $\overleftrightarrow{MN}$ and secant $\overrightarrow{ON}$.
9. Drawing should show ⊙*M* with central ∠*CMR* and tangent $\overleftrightarrow{LR}$.
10. 130° **11.** 130° **12.** 130° **13.** 110° **14.** 70°
15. 70° **16.** 70° **17.** 40° **18.** 70° **19.** 110°
20. Possible answer: $\overline{AC}$, $\overline{BD}$, $\overline{EF}$

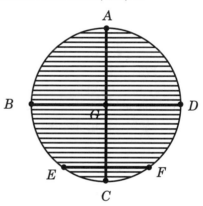

21. Possible answer: ∠*AGB*, ∠*BGC*, ∠*CGD*
22. Possible answer: $\widehat{AB}$, $\widehat{BC}$, $\widehat{CD}$
23. **a.** 90° **b.** 90°
24. No, two arcs cannot have the same degree measurements but different lengths. Since the length of the arc is the same as the degree measurement, if the degree measurement is different, then the two arcs would have different arc lengths. Drawing should show that two arcs with different arc lengths can never equal one another.

6•3 Circumference of a Circle

1. False. The circumference is the distance around the circle.
2. True **3.** 6.2 cm; 3.1 cm **4.** 9.3 cm; 3.1 cm
5. 12.4 cm; 3.1 cm **6.** 15.5 cm; 3.1 cm
7. The ratios all equal 3.1 cm.

8. Hannah is incorrect. The circumference of the circle is $(2)(\pi)(10) = 62.8$ cm.
9. 28.26 m **10.** 25.12 ft **11.** 34.54 in. **12.** 94.2 cm
13. 47.1 ft **14.** 64.37 m **15.** 41.605 ft
16. 118.3152 cm **17.** 82 in. **18.** 56.52 in.
19. If the radius of a circle were doubled, then the circumference of the circle would also be doubled. For example, if the radius equals 2, then the circumference is 12.48. If the radius is doubled to 4, then the circumference is also doubled to 24.96.

6•4 Area of a Circle

1. Jasken is incorrect. Since the area of the square is 16 in², then each side of the square is 4 in., making the diameter of the circle also 4 inches. The area of the circle would be 12.56 in.². The area of the circle is not half the area of the square.
2. 153.86 cm² **3.** 50.24 in.² **4.** 314 m²
5. 706.50 ft² **6.** 38.465 yd² **7.** 211.1336 m²
8. No **9.** Yes **10.** Yes
11. 3 cm; 28.26 cm² **12.** 7 ft; 153.86 ft²
13. 7.5 cm; 176.625 cm² **14.** 5.1 in.; 81.6714 in.²
15. 615.44 cm² **16.** 176.625 ft²
17. 1256 in² **18.** 7850 in² **19.** approximately 6 ft
20. The area of the table is 12.56 ft² or approximately 13 ft². The cost of the table is approximately $390.

6•5 Circumscribed and Inscribed Polygons

1. Drawing should show a triangle circumscribed about ⊙A.
2. Drawing should show a pentagon inscribed in ⊙V.
3. Drawing should show a quadrilateral inscribed in ⊙C.
4. Drawing should show a hexagon circumscribed about ⊙N.
5. The inscribed polygon is a square. The sides of the polygon are all equal.
6. Possible answer: Trevor could fold the square in half and then in half again. He could then unfold the square to find that the center of the square is also the center of the circle.

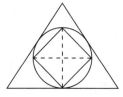

6•6 Measurement of Angles Circumscribed and Inscribed in a Circle

1. Hank is incorrect; the measure of an arc intercepted by a 90° inscribed angle is 180°.
2. 60° **3.** 15° **4.** 70° **5.** 10° **6.** 130° **7.** 39°
8. 52° **9.** 56° **10.** $\overset{\frown}{CE}$ **11.** 100°; 160°; 80°
12. 30° **13.** 78° **14.** 86° **15.** 180° **16.** $\frac{1}{2}$ **17.** 45°
18. 90°; complementary
19. 45°
20. You can find the measurement of the other angle by subtracting the known angle from 360°.

Chapter 6 Review

1. m **2.** h **3.** b **4.** g **5.** c **6.** f **7.** j **8.** d **9.** i
10. k **11.** a **12.** l **13.** e
14. Drawing should show ⊙M with secant $\overleftrightarrow{KL}$.
15. Drawing should show ⊙Q with tangent $\overleftrightarrow{RT}$.
16. Drawing should show ⊙L with central angle KLM and inscribed angle KNM.
17. Drawing should show ⊙F with circumscribed triangle ABC
18. 100° **19.** 130° **20.** 130° **21.** 260° **22.** 80°
23. 50.24 cm; 200.96 cm² **24.** 31.4 in.; 78.5 in.²
25. 60° **26.** 50.24 in.² **27.** 25.12 in.
28. Drawing should include a reconstruction of Tanya's quilt with two circles of radii 3 and 6 inches. The circles are similar in that they both have the same shape. They are different in that the circumference and area of the two circles are different.

Chapter 6 Practice Test

1. Drawing should show ⊙P with diameter $\overline{LT}$ and radius $\overline{PB}$.
2. Drawing should show ⊙M with secant RS and tangent OS.
3. Drawing should show ⊙Q with central angle YQW and chord YW.
4. Drawing should show ⊙T with inscribed angle HIJ and secant IJ.
5. 140°
6. 140°
7. 40°
8. 40°
9. 125.6 cm; 1256 cm²
10. No, since the circular table has an area of 28.26 ft², the 16 ft² tablecloth will not cover the table.

Chapter 6 Test

1.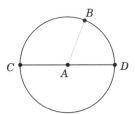

2.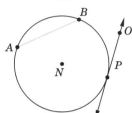

3.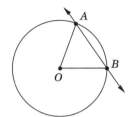

4.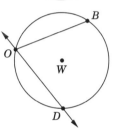

5. 220° **6.** 40° **7.** 250° **8.** 100°

9. Circumference = 75.36 cm; Area = 452.16 cm²

10. Tile B will cover a greater area because its area, 82 square inches, is greater than the area of tile A, which is 78.5 square inches.

Chapter 7   Recognizing Three-Dimensional Shapes

7•1   Properties of Polyhedra

1. The second and fourth shapes should be circled.

2. 8

3. *AFLG, FEKL, EDJK, DCIJ, CBHI, BAGH, ABCDEF, GHIJKL*

4. 18

5. *AB, BC, CD, DE, EF, FA, GH, HI, IJ, JK, KL, LG, AG, BH, CI, DJ, EK, FL*

6. 12

7. *A, B, C, D, E, F, G, H, I, J, K, L*

8. **a.** 5 **b.** 6 **c.** Answers will vary according to the vertex labels.

9. No. A piece of pie cannot be a prism because the crust is curved. A polygon cannot be curved.

10. *XYZW, RSTU; RUWX, STZY; RSYX, UTZW*

11. *RSYX, STZY, UTZW, RUWX, RSTU, XWZY*

12. $\overline{RS}$, $\overline{SY}$, $\overline{YX}$, $\overline{XR}$, $\overline{ST}$, $\overline{TU}$, $\overline{UR}$, $\overline{YZ}$, $\overline{ZW}$, $\overline{WX}$, $\overline{TZ}$, $\overline{UW}$

13. Possible answer: $\overline{RX}$

14. 3 in.

15. **a–d.** Answers will vary according to the label of each vertex.

16. 12

17. 5

18. **a–c.** Answers will vary according to the label of each vertex.

19. No. Since two of the sides must have congruent faces in parallel planes, there are not enough remaining faces to construct a prism.

7•2   Cubes and Rectangular Prisms

1. *JKLM, CDEF; JCFM, KDEL; JCDK, MFEL*

2. Possible answers: *SWGD, TUFE; TSDE, UWGF; TSWU, EDGF*

3. Answers will vary according to the vertex labels.

4. Answers will vary according to the vertex labels.

5. 12

6. 12

7. Answers will vary according to the sizes of the chosen boxes. The length, width, and height of the edges of the rectangular prism should be different. The lengths of the edges of a cube are all equal.

8. Answers will vary according to the size and the shape of the chosen box. You can tell whether the box is a cube or a rectangular prism by measuring the lengths of the edges. If the lengths are all the same, the box is a cube. If the lengths of the edges are different, the box is a rectangular prism.

7•3   Pyramids

1. *ABCD*   **2.** *AEB, BEC, CED, DEA*   **3.** *EF*

4. The pyramid is regular because the base is a square polygon and the lateral faces are congruent isoceles triangles.

5. Answers will vary according to the second chosen variable. However, the first chosen variable should be *E*. Possible answer: *EP*.

6 **a–d.** Answers will vary according to the vertex labels.

7. Answers will vary according to the chosen building and its vertex labels. Answer should include the labeled altitude, slant height, and the reason why the building was shaped as it was.

8. Possible answer: A pyramid and prism are similar in that they both can have rectangular faces for the bases. In addition, all of the lateral faces are

triangular. They are different in that the pyramid only has one base, whereas the prism must have two parallel bases.

7•4   Cylinders and Cones

1. The vertex should be labeled A and the circle should be labeled B.
2. The bases of the cylinder should be labeled C and D.
3. The base should be labeled M and the altitude should be labeled MN.
4. The bases should be labeled R and S and the altitude should be labeled RS.
5. **a.** Answers will vary according to the chosen object and the measurements of the cone. **b.** Possible answer: If the object was an ice-cream cone, then the object would not have a circular base because that is where the ice cream is placed.
6. Answers will vary according to the chosen contents of the containers. Possible answer: Chicken noodle soup could be packed in a cylindrical container. This container is appropriate because cans are stacked on a shelf. Also, the liquid state of the soup allows the soup to take the form of its container. Paper could be packed in rectangular prisms. This container would be appropriate because the container fits the shape of the paper and the containers can easily be stacked on a shelf.

7•5   Spheres

1. The center should be labeled C, and D should be an endpoint on the sphere, forming a radius CD.
2. The center should be labeled M, and the endpoints of the diameter, L and K, should be on the sphere.
3. The center should be labeled R, and T should be an endpoint on the sphere, forming radius RT.
4. The center should be labeled Q, and the points X and Y should be on the sphere, forming diameter XY.
5. 7 m   6. 3 in.   7. 9 ft   8. 7.5 cm   9. 4.5 in.
10. 1.75 m
11. The size of the circle cut through the center is larger than the circle cut anywhere else.
12. Drawing should show a sketch of a globe with a center, a radius, and a diameter. These parts will vary in accordance with the labels chosen. Longitude lines are used to indicate east and west distances.
13. Possible answer: You can find the diameter of a spherical object by first measuring the circumference of the sphere. Since the formula for the circumference of a circle is $C = 2\pi r$, divide the

circumference of the sphere by 2 and then by π. Then, to change the radius into the diameter, multiply the radius by 2.

Chapter 7 Review

1. polyhedron   2. cone   3. rectangular prism
4. sphere   5. prism   6. rectangular pyramid
7. cylinder   8. cube
9. **a.** cube   **b–d.** Answers will vary according to the chosen labels.
10. **a.** sphere   **b–d.** Answers will vary according to the chosen labels.
11. **a.** pyramid   **b–e.** Answers will vary according to the chosen labels.
12. **a.** rectangular prism   **b–f.** Answers will vary according to the chosen labels.
13. **a.** cone   **b–c.** Answers will vary according to the chosen labels.
14. **a.** cylinder   **b–c.** Answers will vary according to the chosen labels.
15. **a.** The diameter of the base of the can must be slightly larger than the diameter of the tennis ball.   **b.** The height of the can must be slightly larger than three times the diameter of the tennis ball.
16. Answers will vary according to the chosen box. If the box is cut apart at the edges one way, there is only one way to tape them back together.

Chapter 7 Practice Test

1. c   2. a   3. e   4. d   5. b
6. Regular pyramid should be labeled with altitude $\overline{RS}$ and slant height $\overline{RQ}$.
7. Sphere should be labeled with center O and radius $\overline{OP}$.
8. Cone should be labeled with base N, altitude $\overline{MN}$, and slant height $\overline{LK}$.
9. Cylinder should be labeled with altitude $\overline{AB}$ and bases A and B.
10. See students' work. Cardboard should be a net of the box.

Chapter 7 Test

1. c   2. e   3. a   4. b   5. d
6. Students' work should show a cone with base circle P, altitude $\overline{MP}$, and slant height $\overline{MK}$.
7. Students' work should show a cylinder with altitude $\overline{XY}$, and bases circle X and circle Y.

8. Students' work should show a regular pyramid with a rectangular base, altitude $\overline{AB}$, and slant height $\overline{AC}$.

9. Drawing should show a sphere with center M, radius $\overline{MP}$, and diameter $\overline{OR}$.

10. a rectangular prism

Chapter 8 Surface Area of Three-Dimensional Shapes

8•1 Surface Area of Rectangular Prisms

1. Josh is correct. Since a rectangular prism has six sides, the lengths of the three sides must each be multiplied by two so that $2(2) + 2(4) + 2(6) = 24$ cm^2.

2. Drawing should show a net rectangular prism with length = 5 cm, width = 4 cm, and height = 2 cm. Surface area = 76 cm^2

3. Drawing should show a net rectangular prism with length = 10 ft, width = 6 ft, and height = 3 ft. Surface area = 216 ft^2

4. Drawing should show a net rectangular prism with length = 8 in., width = 7 in., and height = 6 in. Surface area = 292 in.^{2}

5. Drawing should show a net rectangular prism with length = 2.5 m, width = 1.5 m, and height = 3 m. Surface area = 31.5 m^2

6. c 7. d 8. a 9. b 10. 150 in.^{2} 11. 8.64 cm^2
12. 600 ft^2 13. 1.5 m^2
14. **a.** 54 ft^2 **b.** 2

15. The surface area of Miguel's box is 1,160 in.^{2}. Wrapping paper A has an area of 1,624 in.^{2}. Wrapping paper B has an area of 960 in.^{2}. Wrapping paper C has an area of 1,653 in.^{2}. Wrapping paper B will not work since it will not cover the entire surface area of the box. Wrapping paper C will best fit the requirements of the net of the box.

8•2 Lateral and Surface Area of Cylinders

1. 10π cm 2. 12 cm 3. 226.08 in.^{2} 4. 125.6 m^2
5. 197.82 ft^2 6. 494.55 cm^2
7. Patrick is correct. The surface area = $(2)(3.14)(4)(8) + (2)(3.14)(4^2) = 301.44$ in.^{2}.
8. 879.2 in.^{2} 9. 345.4 m^2 10. 113.04 ft^2
11. 86.35 cm^2 12. 282.6 in.^{2} 13. 226.08 m^2
14. 324.99 ft^2 15. 847.8 cm^2
16. **a.** 11,775 m^2 **b.** 524

17. **a.** If the height of the lateral cylinder is doubled, then the lateral area of the cylinder would also be doubled. For instance, in exercise 3, the lateral area of the cylinder is 226.08 in.^{2}. But if the height is doubled, then the new lateral area, 452.16 in.^{2}, is double the original area. **b.** If the height of the lateral cylinder is doubled, then the surface area of the cylinder would not be doubled. For instance, the cylinder in problem #8 has a surface area of 879.2 in.^{2}. But if the height of the cylinder is doubled, the new surface area is 1,130.4 in.^{2}.

8•3 Surface Area of Rectangular Pyramids

1. c 2. a 3. d 4. b 5. 56 in.^{2} 6. 120 ft^2
7. 175 m^2 8. 96.25 cm^2
9. No, you would also need to know the measurements of the two rectangular bases.
10. 1,279.2 cm^2 11. 299.2 in.^{2} 12. 140 ft^2
13. 139,755 m^2
14. No, you would also need to know the slant heights of the rectangular pyramids in order to find the surface area. (The slant height is not the same as the edge length of the lateral faces.)

8•4 Lateral and Surface Area of Cones

1. c 2. a 3. b 4. d 5. 263.76 cm^2
6. 471 in.^{2} 7. 816.4 ft^2 8. 813.26 m^2
9. Yes; Since half of the diameter equals the radius, all Niki needs to do is divide the diameter by two.
10. 452.16 cm^2 11. 251.2 in.^{2} 12. 942 ft^2
13. 47.1 m^2 14. 417.62 cm^2 15. 785 in.^{2}
16. 1,347.06 ft^2 17. 1,428.7 m^2
18. **a.** 113.04 ft^2 **b.** 395.64 ft^2
19. 94.2 cents
20. The lateral area is not always greater than the area of the base. If the slant height is considerably larger than the radius, then the lateral area will usually be larger than the area of the base. However, if the slant height is considerably smaller than the radius, then the area of the base will usually be larger than the area of the lateral surface.

8•5 Surface Area of Spheres

1. 615.44 ft^2 2. 314 m^2 3. 2,461.76 in.^{2}
4. 1,017.36 mm^2 5. 78.5 m^2 6. 38.465 cm^2
7. **a.** 45.3416 in.^{2} **b.** 17.0416 in.^{2}

8. The surface area of the earth is 16 times larger than the surface area of the moon. For instance, if the diameter of the moon is 10 km, then the surface area is 314 km². Since the diameter of the earth is 4 times larger than that of the moon, then the diameter would be 40 km, and the surface area would be 5,024 km. 5,024 divided by 314 equals 16.

Chapter 8 Review

1. net   2. lateral area   3. surface area
4. 104 cm² 5. 726 in.² 6. 219.8 ft²; 376.8 ft²
7. 16 m²; 20 m² 8. 282.6 ft²; 536.94 ft²
9. 5024 mm²
10. a. 93.5 in.² b. Kamil could cover a total of two cylinders; one that had the measurements of 8.5 inches for the circumference and 11 inches for the height of the cylinder, or one that had the measurements of 11 inches for the circumference of the base and 8.5 inches for the height of the cylinder.
11. 135.648 cm²
12. Answers will vary according to the size of the cardboard and the shapes made. Answers should include surface area of shape, drawing of shape, and the description of how it was made.

Chapter 8 Practice Test

1. 216 cm² 2. 47.1 in.²; 86.35 in.²
3. 48 ft²; 57 ft² 4. 847.8 cm²; 1,554.3 cm²
5. 452.16 m² 6. 251.2 cm²
7. 62 in.² 8. 145,193,600 km²
9. 20.41 in.² 10. 44 ft²

Chapter 8 Test

1. 126.78 in.²; 240.21 in.² 2. 96 cm²
3. 439.6 cm²; 593.46 cm² 4. 6 ft²; 10 ft²
5. 1,256 in.² 6. 42.39 in.² 7. 41.2 ft² left
8. 113.04 in.² 9. 18.84 in.² 10. 94 in.²

Chapter 9   Volume of Three-Dimensional Shapes

9•1   Volume of Rectangular Prisms

1. Garnet is correct. The volume of the box is (6 cm)(4 cm)(5 cm) = 120 cm³.
2. 40 cm³ 3. 360 ft³ 4. 1188 in.³ 5. 35 m³
6. 300 cm³ 7. 140.4 ft³ 8. c 9. d 10. b 11. a
12. 64 cm³ 13. 2,197 in.³ 14. 1,000 ft³

15. 91.125 m³ 16. 12.5 ft³
17. a. 480 ft³ b. approximately 3,600 gallons
18. 15 fish
19. Answers will vary according to the refrigerator measurements.

9•2   Volume of Cylinders

1. Jay is correct. The volume of the can is (3.14)(5 cm)²(10 cm) = 785 cm³.
2. 508.68 in.³ 3. 904.32 ft³ 4. 6,782.4 m³
5. 346.185 cm³ 6. a 7. b 8. c
9. 2,198 in.³ 10. 4,069.44 ft³
11. 30,395.2 mm³ 12. 569.91 m³
13. 282.6 m³ 14. 39.25 ft³
15. a. 235.5 cm³ b. 235.5 g
16. a. Possible answer: The diameter, since it is easier to measure from two points on the circular base rather than from one point on the circle to an inexact center. b. Answers will vary, but the height should be perpendicular to the base. c. Answers will vary, but the units should be cubic units.
17. a. Answers will vary according to object chosen. Possible answer: An oil pipe is shaped like a cylinder because the shape of the container allows the oil to flow more freely. b. Possible answer: Yes. A soda pop can was designed to hold 12 fluid ounces (354 ml). The right size container is important so as to not waste aluminum, which keeps the cost down.

9•3   Volume of Rectangular Pyramids

1. 15 ft³ 2. 20 m³ 3. 54 in.³ 4. 420 cm³
5. $46\frac{2}{3}$ in.³ 6. $108\frac{1}{3}$ m³ 7. 115,651,800 ft³
8. 91,636,272 ft³
9. Six times. Possible answer: If the dimensions of the rectangular prism are $l = 10$ ft, $w = 15$ ft, and $h = 20$ ft, then the area of the prism would be 3000 ft³. The dimensions of the pyramid would be $l = 10$ ft, $w = 15$ ft, $h = 10$ ft for an area of 500 ft³, and 3000 ÷ 500 = 6.

9•4   Volume of Cones

1. 16 m³ 2. 47.1 ft³ 3. 42 ft³ 4. 2,659.58 mm³
5. ≈ 340.17 cm³ 6. 3,282.35 in.³ 7. 150.72 cm³
8. The volume of Cone A is 160 cm³. The volume of Cone B is 640 cm³. The volume of Cone B is four times the volume of Cone A.

9•5 Volume of Spheres

1. 133.04 ft^3 2. $\approx$ 267.95 cm^3 3. 3,052.08 in.^{3}
4. $\approx$ 2,143.57 mm^3 5. 696.56 m^3 6. 8,177.08 in.^{3}
7. 14,130 ft^3 8. $\approx$ 1,022.14 m^3 9. 452.16 ft^3
10. No. Possible answer: If the radius sas three times as long, the volume would be 27 times as great.

9•6 Distinguishing Between Surface Area and Volume

1. surface area
2. volume
3. volume
4. surface area
5. volume
6. surface area
7. volume
8. a. 91.75 ft^2
   b. 60 ft^3
9. a. 37.68 in.^{2}
   b. 28.26 in.^{3}
10. Square meters measures area. Cubic meters measures volume. Possible answer: Square meters are used to measure the amount of carpet needed to carpet a room, and cubic meters are used to measure the amount of space in an elevator.

Chapter 9 Review

1. b 2. c 3. e 4. a 5. d
6. 120 ft^3 7. 729 cm^3 8. 791.28 in.^{3} 9. 10 m^3

10. 200.96 ft^3 11. $\approx$ 5,572.45 cm^3
12. a. 4,500 cm^3
    b. 10 cm^3
    c. 2,250 cm^3
    d. 2,925 cm^3
13. a. 22,608 in.^{2}
    b. 2,543.4 in.^{3}
14. Answers will vary according to the box designs. Drawing should include a plan for the box, along with the dimensions and a sketch.

Chapter 9 Practice Test

1. 168 cm^3 2. 1,157.625 m^3 3. 1,808.64 ft^3
4. 224 in.^{3} 5. 307.72 cm^3 6. $\approx$11,488.21 in.^{3}
7. 1,884 ft^3 8. 392.5 cm^3 9. $523\frac{1}{3}$ cm^3
10. Yes, since the volume of the cone is less than the volume of the ice cream, the cone would overflow if the ice cream were to melt.

Chapter 9 Test

1. 255 in.^{3}
2. 1,331 cm^3
3. 602.88 m^3
4. 108 in.^{3} 5. 425.08 cm^3 6. 38.77 in.^{3}
7. 38.86 in.^{3} 8. 84.78 cm^3 9. 18.84 cm^3
10. Yes. Since the whole cup holds 84.78 cm^3 of water and the cup filled to the line holds 18.84 cm^3 of water, there is enough room for only 65.94 cm^3 of water after the cup is filled to the line. Therefore, pouring 70 cm^3 of water would cause the cup to overflow.